Springer Series in Information Sciences   34

T0189784

Springer
Berlin
Heidelberg
New York
Barcelona
Hong Kong
London
Milan
Paris
Singapore
Tokyo

Physics and Astronomy   ONLINE LIBRARY

http://www.springer.de/phys/

Springer Series in Information Sciences

Editors: Thomas S. Huang   Teuvo Kohonen   Manfred R. Schroeder

Armin Gruen   Thomas S. Huang (Eds.)

Calibration and Orientation of Cameras in Computer Vision

With 77 Figures and 25 Tables

 Springer

Professor Armin Gruen

ETH-Hönggerberg
Institut für Geodäsie und Photogrammetrie
8093 Zürich, SWITZERLAND

Professor Thomas S. Huang

University of Illinois
2039 Beckman Institute, 405 N. Mathews
Urbana, IL 61801, USA

Series Editors:
Professor Thomas S. Huang

University of Illinois, 2039 Beckman Institute, 405 N. Mathews,
Urbana, IL 61801, USA

Professor Teuvo Kohonen

Helsinki University of Technology, Neural Networks Research Centre
P.O. Box 5400,
02150 HUT, Espoo, FINLAND

Professor Dr. Manfred R. Schroeder

Drittes Physikalisches Institut, Universität Göttingen, Bürgerstrasse 42-44
37073 Göttingen, GERMANY

ISSN 0720-678X
ISBN 978-3-642-08463-8

Library of Congress Cataloging-in-Publication Data

Calibration and orientation of cameras in computer vision /
Armin Gruen, Thomas S. Huang (eds.).
p. cm. – (Springer series in information sciences, ISSN 0720-678X; 34)
Includes bibliographical references.
1. Computer vision. 2. Photogrammetry. 3. Cameras–Calibration.
I. Gruen, A. (Armin) II. Huang, Thomas S., 1936– III. Series.
TA1634.C35 2001 621.39'9–dc21 2001020735

Springer-Verlag Berlin Heidelberg New York
a member of BertelsmannSpringer Science+Business Media GmbH

http://www.springer.de

© Springer-Verlag Berlin Heidelberg 2010
Printed in Germany

Cover design: *design & production* GmbH, Heidelberg

Preface

This book was conceived during the Workshop "Calibration and Orientation of Cameras in Computer Vision" at the XVIIth Congress of the ISPRS (International Society of Photogrammetry and Remote Sensing), in July 1992 in Washington, D.C. The goal of this workshop was to bring photogrammetry and computer vision experts together in order to exchange ideas, concepts and approaches in camera calibration and orientation. These topics have been addressed in photogrammetry research for a long time, starting in the second half of the 19th century. Over the years standard procedures have been developed and implemented, in particular for metric cameras, such that in the photogrammetric community such issues were considered as solved problems. With the increased use of non-metric cameras (in photogrammetry they are revealingly called "amateur" cameras), especially CCD cameras, and the exciting possibilities of acquiring long image sequences quite effortlessly and processing image data automatically, online and even in real-time, the need to take a new and fresh look at various calibration and orientation issues became obvious. Here most activities emerged through the computer vision community, which was somewhat unaware as to what had already been achieved in photogrammetry. On the other hand, photogrammetrists seemed to ignore the new and interesting studies, in particular on the problems of orientation, that were being performed by computer vision experts. And it seems that even nowadays the main interest that photogrammetrists take in the work of computer vision scientists is in the areas of image analysis and scene understanding.

For various reasons it took almost a decade to finally put this book together and publish it. Nevertheless, the content of this book is still relevant. Since 1992 there has been of course some progress in the understanding and solving of orientation problems and also in calibration, but we hope to have captured at least some of these developments through a recent chapter updating cycle.

We have brought together internationally renowned authors, both from photogrammetry and computer vision, all of them experts in their respective fields. We present in total seven chapters, with four contributions from photogrammetrists and three by computer vision experts. This is not a textbook. Therefore, it is neither consistent in diction nor in content. It is not complete in the coverage of all interesting issues and it has certain overlaps. This is intended and simply reflects the current situation in the research fields of photogrammetry and computer vision.

We hope that this book represents interesting study material for students, researchers, developers and practitioners alike. If readers from photogrammetry and computer vision find the material equally rewarding then one of our original goals is achieved: To bring both communities together on the basis of some very exciting scientific and technical subjects.

It is with great pleasure and respect that I thank the authors for their stimulating contributions and enduring patience and the publisher for not giving up on this book over the many years of preparation.

Zürich, March 2001   *Armin Gruen*

Contents

List of Contributors

Horst A. Beyer
IMETRIC SA
2900 Porrentruy
Switzerland
horst.beyer@imetric.com

Olivier D. Faugeras
I.N.R.I.A.
2004 route des Lucioles
B.P. 93 06902 Sophia-Antipolis
France
luong,faugeras@sophia.inria.fr

Wolfgang Förstner
Institut für Photogrammetrie
Universität Bonn
Nussallee 15
53115 Bonn
Germany
wf@ipb.uni-bonn.de

Clive S. Fraser
Department of Geomatics
University of Melbourne
Parkville Victoria 3052
Australia
c.fraser@eng.unimelb.edu.au

Donald B. Gennery
Jet Propulsion Laboratory
California Institute of Technology
4800 Oak Grove Drive
Pasadena, CA 91 109-8099
USA
gennery@robotics.jpl.nasa.gov

Armin Gruen
Institute of Geodesy
and Photogrammetry
ETH-Hoenggerberg
8093, Zuerich, Switzerland
agruen@geod.baug.ethz.ch

Quang-Tuan Luong
AI Center
333 Ravenswood av
Menlo Park, CA 94025, USA
luong@ai.sri.com

Steve Shafer
Microsoft Corp.
One Microsoft Way
Redmond, WA 98052, USA
stevensh@microsoft.com

Reg G. Willson
3M Engineering Systems
3M Center, Building 518-1-01
St. Paul, MN 55144
USA
rgwillson@mmm.com

Bernhard P. Wrobel
Institute of Photogrammetry
and Cartography
Darmstadt University of Technology
Petersenstrasse 13
64287 Darmstadt
Germany
wrobel@gauss.phgr.verm.-
tu-darmstadt.de

1   Introduction

Armin Gruen

Calibration and orientation of cameras and images are procedures of fundamental importance for quantitative image analysis (IA). Essentially, quantitative IA is concerned with the identification of information (image primitives,
features, objects) in image space and its transfer and representation in object
space. With some simplification, qualitative IA deals with image interpretation by image understanding, that is with the problem of *what* is imaged in
one or more images and with associated search and recognition strategies.
Quantitative IA on the other hand addresses the question of *where* and *how*
an imaged object is located in object space. Calibration and orientation, at
first sight, seem only to be of interest in the context of quantitative IA. Obviously, basic component algorithms of IA, like image matching, segmentation,
feature extraction, etc. are favourably supported and constrained by the use
of orientation parameters. However, the use of orientation parameters may
be important for qualitative IA as well. They may support the image understanding procedures in the sense that their knowledge reduces the size of the
solution spaces and the number of admissible hypotheses, and provides for
more reliable results.

In the case where object location and orientation in 3-D space has to
be inferred from 2-D image data, knowledge of the camera model (full perspective frame, linear array scanner, point scanner, or others), that is the
geometric and radiometric laws of image formation, is indispensable. This
book deals exclusively with the full perspective camera model, although it
is acknowledged that, in particular, linear array scanner models are gaining
quickly in relevance. Linear array scanners have been used in close-range applications for quite a while, on satellite platforms for about 25 years, and are
currently at the threshold of operational use in airborne systems.

The terms "calibration" and "orientation" were introduced and defined in
photogrammetry a long time ago. In computer vision deviating definitions are
sometimes used. For clarification we will reiterate here the photogrammetric
definitions of some technical terms:

Calibration. Determination of the parameters of the interior orientation
of a camera or an image. These are the coordinates of the principal point
and the camera (or image) constant. Please note that the camera constant
differs in general from the focal length. Some authors will also include the
parameters of the symmetric radial lens distortion in the set of parameters
for the interior orientation.

Springer Series in Information Sciences, Vol. 34
Calibration and Orientation of Cameras in Computer Vision
Eds.: Gruen, Huang   © Springer-Verlag Berlin Heidelberg 2001

Orientation. Split into interior (inner) and exterior (outer) orientation. The exterior orientation describes the three parameters of the object coordinates of the perspective center (to be precise, the perspective center of a camera coincides with the object-sided image of the center of the lens aperture, the "entrance pupil") and the three rotational parameters. The orientation may be performed for single images, image pairs or multiple image configurations.

System Calibration. Determination of the interior orientation parameters and all systematic errors (defined as physical deviations from the mathematical camera model) of all cameras (images) used in a particular project.

Self-Calibration. Determination of all systematic errors (plus, possibly, the parameters of the interior orientation) simultaneously with all other system parameters (e.g. of the bundle adjustment) using the concept of additional parameter estimation. Thus, self-calibration is just one particular technique for estimating and compensating systematic errors (parameters of the interior orientation, introduced with incorrect values, will also act as systematic errors), among many others. Therefore, system calibration may be done using self-calibration.

The early theories and concepts for orientation were laid down during the second half of the 19th century and in the 1930s and 1940s. Most of these publications were written in German and until nowadays they were only accessible to a minority of researchers. Thus, it is not surprising that the computer vision literature is filled with contributions which are, to say the least, a bit outdated. Therefore, it is a very deserving task not only to shed some light on the current state of research in orientation but also to give proper credit to those who made significant findings some generations ago.

The second chapter of this book, written by B. Wrobel on "Minimum Solutions for Orientation", gives useful hints and a collection of early literature on the problem of orientation. Beyond that, new relevant research results, most of which have been generated by the computer vision community, are also addressed.

The emphasis is on minimum solutions (that is "minimum information solutions"), uniqueness conditions and cases of degeneracy. The focus is on the so-called basic orientation tasks, 2D–2D (e.g. relative orientation), 2D–3D (e.g. spatial resection) and 3D–3D (absolute orientation). Among the information elements used in orientation (points, lines, surfaces, digital picture functions) only points are covered here, since there are still many gaps in the proper understanding of the functioning and effects of the other elements. It is interesting to note that, under minimum conditions, for all three problems (2D–2D, 2D–3D and 3D–3D) closed-form direct solutions for point primitives are available.

B. Wrobel's contribution includes a wealth of useful information. It shows clearly that the research activity in this area originated in the early years of projective geometry and photogrammetry and was picked up again in the

1980s and 1990s by the computer vision community. Today, there is a great
number of procedures and algorithms available, and choosing among them is
not always a simple task.

W. Förstner's contribution on "Generic Estimation Procedures for Orientation with Minimum and Redundant Information" represents a natural
and necessary extension of the previous chapter into the domain of redundant observations and estimation. The author summarizes nicely some of the
concepts that have been developed in photogrammetry, and to a lesser extent in computer vision over the past 30 years. In calibration and orientation
four major tasks have to be addressed, all of which are of reasonable complexity: camera modelling, 3D geometry ("network design"), error handling
(including quality control) and full automation. The geometric part of camera
modelling, including deviations from the pinhole model, is well established
today and, together with other advancements, allows for very high accuracy
results. This has finally paved the way for the successful use of photogrammetric techniques in a great variety of diverse applications. The radiometric
part of the camera model, however, has been widely neglected in the past.

Network design in photogrammetry is based both on sound theories, as
in aerial photogrammetry, and on experience, trial-and-error and simulations
at close-range. In computer vision much progress has been achieved lately in
the study of the properties of close-range systems.

In photogrammetry a theory of error definition, handling and compensation has emerged over the years, based on well-proven statistical concepts of
parameter and interval estimation. While one particular classification scheme
distinguishes the error types as data errors, model errors and design errors,
another is based on the notions of random errors, systematic errors and gross
errors. These errors may appear both in the functional and the stochastic
parts of the estimation models.

Effective quality control is a very important issue, in particular in fully
automated systems. Here the concepts of system redundancy and reliability
structure are crucial for the detectability and locatability (please note the
difference!) of errors. In summary, this chapter shows that estimation and
analysis procedures and, to a lesser extent, design knowledge are available,
and these are very helpful in orientation procedures as well as in object
reconstruction tasks.

C. Fraser's chapter on "Photogrammetric Camera Component Calibration: A Review of Analytical Techniques" touches on another crucial issue.
Calibration, in whatever form, is an essential part of the establishment of a
sophisticated camera model and only through advanced and complete calibration can we expect to obtain a high metric accuracy system performance
(relative accuracies of up to 1:1 000 000 have been reported).

This chapter deals in a tutorial fashion with the well-investigated major systematic errors that may appear in close-range systems and lead to a
departure from the collinearity model: Symmetric radial distortion, decen-

tering distortion, focal plane unflatness and focal plane distortion. From the many possible calibration procedures, test-range calibration, self-calibration and plumbline calibration are addressed in more detail.

D.B. Gennery presents the issue of camera calibration, including symmetrical radial lens distortion in more procedural detail ("Least-Squares Camera Calibration Including Lens Distortion and Automatic Editing of Calibration Points"). In particular, he introduces a refinement to the analytical formulation of the lens distortion by considering the non-perpendicularity of the imaging plane to the optical axis. He addresses aspects of the least-squares solution and the procedure of rejection of outliers in the measurements of image points.

R.G. Willson and S. Shafer ("Modelling and Calibration of Variable-Parameter Camera Systems") extend the concept of calibration further in order to be able to also deal with variable-parameter camera systems, that is with cameras which can be focussed and zoomed. It is well known that focussing changes the camera constant, the location of the principal point and the lens distortion, while zooming changes the focal length and also the other parameters. Thus, it becomes a formidable task to calibrate cameras whose parameters are actively modified during picture taking in a project. The authors present a new general model for the image formation process, including both a geometric and a radiometric part, involving more calibration parameters than the traditional approach. The new model is presented on a conceptional level. It will be interesting to see how it can be implemeted in a real-world environment.

A. Gruen and H.A. Beyer ("System Calibration through Self-Calibration") present the concept of self-calibration as it is used in photogrammetry since the early 1970s. Starting with the notion of a general bundle model it is shown how all cases of metric camera bundle, spatial resection and spatial intersection derive from this general concept. Each procedure can be executed in the self-calibration mode and least squares is the standard parameter estimation approach. Additional parameters are used in order to model the deviation from the pinhole model. A sufficiently large number of parameters leads to high model fidelity, but triggers the danger of overparameterization. Therefore, a reliable procedure for stability control of these parameters is crucial. Procedural details depend on whether the purpose of self-calibration is optimum point positioning in object space, estimation of the exterior orientation elements (important for egomotion and navigation), or investigation of a system's systematic errors. As a more tutorial-type introduction the determinability of standard self-calibration parameters (including interior orientation) under various network conditions is presented. This gives clear indications about the network configurations for which certain parameters are determinable or not.

In the last chapter Q.T. Luong and O.D. Faugeras demonstrate with a concrete vision task ("Self-Calibration of a Stereo Rig from Unknown Camera

Motions and Point Correspondences") the computation of both the intrinsic and extrinsic camera parameters of a moving stereo rig. They show theoretically and with experiments on real images that it is possible to calibrate a stereo rig by pointing it at the environment, selecting points of interest automatically and tracking them independently in each image while moving the stereo rig with an unknown motion. In this context the fundamental matrix is computed using a robust method and the intrinsic parameters are obtained through the Kruppa equations arising from epipolar geometry.

In summary, these contributions, despite the use of partly different notations in photogrammetry and computer vision, show very well the current status of calibration and orientation techniques. Mathematically, the geometrical conditions of point-based systems are fairly well understood. Today the great challenge lies in the investigation of line- and surface-based approaches and of hybrid systems. Other issues which are of great concern, in particular in non-metric camera-based and in automated image analysis systems, are the modelling of systematic errors and the detection, location and removal of blunders of various types. Also, the awareness for the necessity of sophisticated network design and system analysis techniques should be sharpened. Furthermore, at least within the photogrammetric community, the need to develop and investigate sensor models for imaging and recording geometries other than perspective, as they arise for linear array scanners, microwave sensors, laser scanners and the like, are topics of great relevance. Finally, the issues of calibration, orientation, and the related topics of sensor and error modelling, network design and system quality control will stay with us for some time and will continue to provide a multitude of interesting research topics.

2 Minimum Solutions for Orientation

Bernhard P. Wrobel

Summary

A review will be given on minimum solutions of various orientation tasks of photogrammetry and computer vision. This review is based on examining the latest developments and also by looking back at the early beginnings of orientation procedures of photogrammetry to demonstrate that there are already answers to a lot of the conceptual questions of today. The first part of the chapter is devoted to the historical roots of orientation approaches and to a comparison of two very popular camera models: The first is based on projective collineation and the second on perspective collineation. In the next part of this chapter a systematic overview on standard orientation tasks of photogrammetry and computer vision, from single image to multiple image orientation, is given. Next, among the orientation tasks using corresponding points three basic ones, 2D-2D relative orientation (also called recovery of motion), 2D-3D image orientation (or space resection or camera calibration) and 3D-3D absolute orientation (3D similarity transformation) are discussed in detail. For all of them iterative and general closed-form direct solution procedures are available although at different levels of compactness.

Due to the importance for automation of orientation particular attention will be focussed on the representation of rotation, free from singularities in full parameter space, and on the critical configurations of points in 3D object space, i.e. center(s) of perspective and points on the object surface. Existence and uniqueness of orientation solutions directly depend on both. Complete uniqueness conditions of the three basic orientation tasks have already been derived long ago, as part of basic research work in projective geometry, namely in the 19th century and in 1930–1948. Thus the importance of projective geometry for vision geometry, not only from its beginnings for photogrammetry, but also today for new vision tasks, cannot be overestimated.

2.1 Introduction

The orientation of images in object space is a prerequisite for image evaluation. As such it is often regarded as a troublesome impediment prior to the proper task of deriving object space information from image data. However,

Springer Series in Information Sciences, Vol. 34
Calibration and Orientation of Cameras in Computer Vision
Eds.: Gruen, Huang © Springer-Verlag Berlin Heidelberg 2001

after a short time you will find out that the problems of orientation entail interesting mathematics, closely related to applications. Among the many fascinating aspects are:

- Orientation proves to be a new task and a very old task.
- It may be very simple or mathematically nontrivial.
- The orientation approach may be represented by a projective camera model or by the more physically based perspective camera model and may or may not include additional parameters for calibration.
- The parameters of an orientation approach are usually computed from corresponding features of image space and object space. Various features are in use or still being researched: traditionally points or lines, but also other strong features or simply the original grey values themselves.
- There are numerous numerical processes of orientation, many of which are iterative; however in recent years powerful direct procedures have been developed.
- Uniqueness or multiplicity of solutions, also cases of degeneracy, are very fascinating and demand a deep insight into projective geometry.

For an introduction we consider some of the above points in more detail. The orientation of images is clearly involved in tasks of current interest such as the control of autonomous vehicles or robots, the quality assessment of small or very large industrial products (Fig. 2.1), surface monitoring of historical monuments or of human bodies in medicine, product advertising together with multi media or realistic 3D models for virtual and augmented real environments, and so forth. Orientation of images as such has therefore attracted much interest in computer vision research. When preparing this chapter I became aware how many new and outstanding contributions to

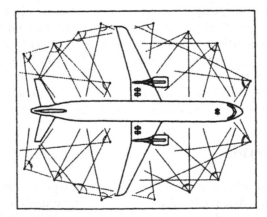

Fig. 2.1. Image configuration for the final measurement of the aircraft Airbus A 321, 1993. (With permission of Deutsche Aerospace Airbus, Hamburg, and Institut für Photogrammetrie und Bildverarbeitung, Technische Universität Braunschweig.)

Fig. 2.2. Aimé Laussedat's phototheodolite with devices for direct measurement of some orientation data (1859). (Reprinted, with permission, from the *Manual of Photogrammetry*, 4th ed., p. 6, copyright 1980, by the American Society for Photogrammetry and Remote Sensing).

orientation have been given in recent years, see e.g. the *IEEE Transactions PAMI*, other journals or conference proceedings of computer vision; also the references in Chap. 8.

On the other hand the early approaches to orientation centuries ago are with the measurement instruments of spherical astronomy, of geodesy (surveying) and of photogrammetry. Many of the findings from early times have been forgotten or not been noticed by the computer vision community. For example in 1795 J.L. Lagrange [1] the famous mathematician, gave a polynomial solution of space resection which is a main part of camera orientation. Shortly after the invention of applicable photography in 1837, photogrammetry came into existence. In France A. Laussedat, one of the first photogrammetrists of the world, used a combination of a camera and theodolite for taking images on earth (Fig. 2.2) in 1859. The orientation in object space with that equipment is very simple, in principle by directly measuring the orientation data.

la photographie aérienne

Fig. 2.3. First air photography from balloons in France since 1858. (Reprinted, with permission, from Kodak-Pathé "la photographie aérienne". Lettre de Paris, no. 23, 1969).

Only a few years later balloons (Fig. 2.3) were used as platforms for aerial photography (G.F. Tournachon 1858), also for photogrammetry. From that time on the orientation of images taken by moving cameras has been stated as a nontrivial mathematical problem, since it has to be solved indirectly from relationships between corresponding features of image space and object space.

In Germany in the second half of the 19th century and later also in Austria some mathematicians laid down the mathematical foundations of photogrammetry. They defined many terms still in use today, e.g. interior and exterior orientation, relative and absolute orientation. An early and prominent representative is Sebastian Finsterwalder 1899 [2,3] from the Munich University of Technology. With a stereo pair of photographs (from the village of Gars at the River Inn) taken from a balloon Finsterwalder presented a mathematically exact solution of the orientation and of reconstruction of the terrain (a topographic map with contour lines) in 1903. Contributions to photogrammetry from universities including the intrinsic details of orientation (number of solutions, their failure and instability cases on so-called 'dangerous surfaces' ('gefährliche Flächen' [2]) have been thoroughly investigated and solved, mainly by Austrian mathematicians from the Vienna University of Technology [4–6]. New and supplementary contributions have been presented by the photogrammetrists R. Bosshardt 1933, R. Finsterwalder 1934, E. Gotthardt 1939 [7], K. Killian 1945 [8] and others. For references and details about that period see the excellent surveys by W. Hofmann [9] and Hohenberg and Tschupik [10]. The indirect access to orientation – in the past indispensible with aerial photography – perhaps has already become obsolete. Today orientation data may be obtained directly and rather easily by using modern navigational sensors with satellites as for instance the NAVSTAR Global Positioning System for the position of a camera (Fig. 2.4) and an Inertial Navigation System (INS) for its attitude. Such techniques, however, can hardly be used in close-range photogrammetry applications, see Fig. 2.1 and the new computer vision tasks mentioned above where instead fully automized orientation by image processing are very important.

I think the complexity of the new computer vision tasks is quite well represented by the following example: Let us take a CCD video camera, certainly equipped with autofocus and zoom optics (i.e. the interior orientation of the camera is not fixed), and let us capture a large sequence of overlapping digital images of an object, a building say, just by walking around it. This work is quickly and easily finished, however, which are the approaches and procedures to orient all these images automatically in an acceptable time and yielding prespecified quality features? The activities of computer vision and of photogrammetry move quickly to solve tasks like these. We will see that the orientation procedures – being one part of such objectives – are already on a high level of investigation.

Table 2.1. Orientation approach and its parameters – a comparison of two models.

Camera model based on **projective collineation**	Camera model based on **perspective collineation** (see Fig. 2.5)
$$\underset{(3\times 1)}{\boldsymbol{p}_i^{'*}} = \underset{(3\times 4)}{\boldsymbol{A}} \cdot \underset{(4\times 1)}{\boldsymbol{p}_i^*}, \quad i=1,\dots \qquad (2.1)$$	$$\underset{(3\times 1)}{\boldsymbol{p}_i^{'}} = \lambda_i \cdot \underset{(3\times 3)}{\boldsymbol{R}} \cdot \underset{(3\times 1)}{\boldsymbol{p}_i} + \underset{(3\times 1)}{\boldsymbol{t}}, \quad i=1,\dots \qquad (2.2)$$ $$\boldsymbol{R}^T \boldsymbol{R} = \boldsymbol{I}$$
linear relationship	nonlinear relationship
$\boldsymbol{p}_i^{'*}, \boldsymbol{p}_i^*$ projective coordinates of point P_i in image space and object space, resp.	$\boldsymbol{p}_i^{'}, \boldsymbol{p}_i$ cartesian coordinates of point P_i in image space and object space, resp. λ_i scale factor of vector $\boldsymbol{p}_i^{'}$
11 relevant parameters of matrix $\boldsymbol{A}$ per image	exterior orientation: 6 parameters per image $\boldsymbol{R} = \boldsymbol{R}(\omega, \phi, \kappa)$ rotation angles $\boldsymbol{t} = (X_0, Y_0, Z_0)^T$ object space coordinates of center of perspective
	interior orientation: $(x_0, y_0, -c)^T$ image coordinates of principal point H' c camera constant $\Delta x_i(x,y), \Delta y_i(x,y)$ correction functions of image coordinates x_i, y_i
variable focus or zoom optics accepted	stable camera and stable optics required
many correspondences needed (six corresp. points for one image)	not so many correspondences needed (three corresp. points for one image)
low model stability (Sect. 2.4.4) eq. (2.1) needs normalization	high model stability (Sect. 2.4.4)
corresponding points on an object plane lead to a critical configuration	corresponding points on an object plane lead to a stable configuration
The 11 relevant elements of matrix $\boldsymbol{A}$ correspond to six parameters of exterior orientation, and five of interior orientation including two for an affine image correction function $\Delta x(x,y), \Delta y(x,y)$	

Fig. 2.4. Direct measurement of the space position of an aerial camera using the NAVSTAR Global Positioning System (in a kinematic mode with two systems: one aboard the aircraft, one on earth), 1990. (Reprinted, with permission, of ZEISS Germany.)

Orientation procedures are numerous. A fundamental criterion for grouping is the camera model used, namely the projective camera model and the perspective camera model, see Table 2.1, Fig. 2.5 and equations (7.1), (7.2) and (7.7), (7.8) in Chap. 7.

Both models have been discussed in photogrammetry from its beginnings [2]. The projective approach, also called DLT (= Direct Linear Transformation, Abdel-Aziz and Karara 1971 [11]) is popular again due to its linearity and easy handling. DLT has been regarded as being attractive primarily for general-purpose nonphotogrammetric cameras, being equipped with variable focus and zoom optics, e.g. reflex cameras, also CCD video cameras from the consumer market, which are widespread in computer vision, see [12].

Formulae for transformation of the parameters from one model to the other are given. Melen [14,15] has presented a strict and unique decomposition procedure of the DLT matrix A into all eleven parameters of the corresponding perspective camera model without any assumption on one of them. For that Förstner has found very simple formulae [16]. A similar decomposition

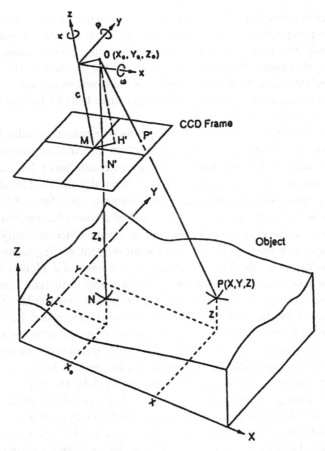

Fig. 2.5. Perspective collineation and its parameters (see Table 2.1).

has been derived for the case that all reference points are lying on one object plane and with matrix A reducing to a 3×3 matrix. Then, from the eight relevant parameters of this matrix A the eleven parameters of the perspective camera model (Table 2.1) can only be determined if three are supposed to be given a priori. Melen [14] investigates this transformation with different sets of a priori given parameters and concludes that in practice only with a given focal length and position of the principal point of the camera can its remaining parameters be computed with sufficient numerical accuracy.

Fuchs [17] applies projective geometry for compact linear formulation of orientation (single image orientation, relative orientation of two images and orientation of many overlapping images). Haggrén and Niini [18] use the projective approach for relative orientation and compute normalized images with optical axes strictly parallel and normal to the baseline. Brandstätter [19–21] building on previous projective work of Rinner [22] and Thompson [23] updates it to a very short linear eight-point solution of relative orientation.

He also gives relationships for a consistent parameter transformation from projective to perspective model parameters. The mathematician von Sanden [24] apparently has been the first to present the linear eight-point solution in 1908, also discussing degenerate cases in detail. A similar treatment of relative orientation has been elaborated in a tremendous amount of work in the computer vision community approximately since 1981 by Longuet-Higgins [25] and many others, see Chap. 8.

However, despite the benefits of the projective camera model there are some drawbacks: less stability, more parameters and more correspondences are involved; also, it is complicated to deal with nonlinear lens distortion which can amount to rather large values (eventually up to $\pm 60\,\mu$m in the image plane), especially with standard CCD cameras, see Table 2.1. The low stability can be improved in part by the classical means of normalization, i.e. by translation of the corresponding points in image space and object space into their respective mean positions and by subsequent scaling to roughly the same coordinate intervals. The instability in the case of a plane object has to be regarded as a real disadvantage, since in many industrial applications object planes have to be expected. Fortunately, this particular instability may be removed by the projective camera model if specialized for the plane case: the 2D projective – 2D object space transformation with the matrix A of Table 2.1 now being a 3×3 matrix. Then only four corresponding points are needed and indeed a stable orientation is obtained. It has to be stressed, however, that image evaluation has to be restricted to the plane containing the given orientation points (reference points). With general work in 3D space a mixed 2D/3D object surface within one image frame may often occur. Even with a statistically optimal decision for the "best model" as a compromise, 2D or 3D, mixed object surfaces will remain to be a source for errors of image evaluation. Of course, this inherent weakness of the projective camera model will show up again in all the other orientation tasks with more than one image. For more on stability see Sects. 2.3.1 and 2.4.

Usually, in photogrammetry rather stable cameras are used, which retain their interior orientation including refined calibration functions $\Delta x(x,y)$ and $\Delta y(x,y)$ over a long period of time or at least over one long sequence of images of a job. Physically based sophisticated calibration (with a testfield or multi-image on-the-job-calibration) are worthwhile and can be refined in a natural way to very high precision with the perspective model and with full statistical support. In recent time this opinion has also been shared by computer vision workers, see Weng et al. [26], p. 967, who states that real-time calibration is not necessary in most applications. However, one clear distinction should be kept in mind throughout this chapter: The computation of exterior orientation parameters is what is here meant by orientation, whereas the inclusion of one or more parameters of interior orientation is already calibration, for which I refer the reader to the chapters of the other authors of this book. By the way, it is easier to simulate the projective approach with refined perspective equations than vice versa. I think these are the main reasons for the preference of this model in photogrammetry. So, it enables us to meet flexibly its broad-band evaluation and mensuration tasks from very simple and low accuracy tasks to complex and high precision tasks with only one camera model for all types of object surfaces. It should be kept in mind that results of photogrammetry are always defined in 3D metric space.

Nevertheless, there are tasks of image processing for which other criteria have to be given priority, e.g. very short processing time and reconstruction of object points not in 3D metric space – as we are used to. I mention the real-time guidance of a robot to touch or circumvent objects in 3D space. This problem has not necessarily to be modeled in 3D metric space, it can be formulated as well and more simply in projective or affine 3D space as has been proposed by Faugeras [27,28]; Maybank and Faugeras [29]; Shashua and Nabab [30]; Ninii [31,32]; and Quan [33]. For that the projective camera model seems to be optimal. Orientation with the projective approach is given in other chapters of this book.

In this chapter orientation procedures with the perspective camera model are discussed and we only consider corresponding points from image space and object space as given information. For this type of feature orientation procedures exist with the longest tradition and are best investigated. Therefore, references to the early contributions from projective geometry and from photogrammetry are also mentioned. We restrict still further this presentation primarily to orientation at minimum information. Of course, this is very important if actually only a minimum of information is available, but in the case of redundant and very noisy correspondences a RANSAC-like technique for outlier detection may be applied for which the minimum solution has to be known as well, see Fischler and Bolles [38] and Chap. 3.

The remainder of this chapter is structured as follows. We first look for the standard orientation tasks that can be identified in photogrammetric or computer vision problems if one or more images have to be evaluated (Sect. 2.2). Then for the most important basic orientation tasks algorithms are selected, preferably direct procedures, which means the solution is computed with an a priori fixed number of arithmetic operations (Sect. 2.3). In Sect. 2.4 from the literature detailed conditions are compiled for which the solution exists uniquely, will be multiple or will be indeterminate or unstable (the problem of so-called "dangerous surfaces"). Finally, we sum up where we are and what is still missing for optimal orientation of cameras of today (Sect. 2.5).

2.2   Standard Orientation Tasks of Photogrammetry and Computer Vision

Orientation determination is always integrated into image evaluation for a specific application. To obtain a platform for an overview away from details of applications the following formal principle for ordering has been adopted: All combinations have been set up from 2D images and 3D object space coordinate systems which are linked by corresponding features (Table 2.2). We start with two sets (see no. 1–3), then three (no. 4–7), four (no. 8–11) and finally more than four (no. 12, 13). Of these combinations we put aside those which are not reasonable or which have not been found in any application. The others are checked for identification of basic orientation tasks. The computational approaches of these tasks are presented in Sect. 2.3. Points and straight lines are natural elements of projective geometry. They have been considered as corresponding features for orientation from the beginnings of photogrammetry [2,3]. Lines were used in theoretical or experimental investigations but rarely in commercial applications.

The situation has changed completely since the advent of digital image processing. Fully automated orientation of images is one of the research areas of computer vision (see this book and [28], [34]) and with a little delay also of photogrammetry (see the review by Strunz [35], also [36,37]). For definition and computation of salient features we refer to standard procedures of image processing (e.g. [34]). Mostly, features with a geometric meaning about the object surface have been investigated, such as points, lines (straight or curved, Schenk and Toth [39]), corners, ... and surfaces (Ebner and Müller [40], Rosenholm and Torlegård [41]). Schickler [42] reports single image orientation through corresponding lines, now introduced into a routine for fully automated orthoimage production in North-Rhine–Westphalia (Germany). Besides geometrically motivated features also "interesting texture patches", good for a strong least-squares match, have been successfully introduced for orientation [43].

In 1987 new approaches were presented for object surface reconstruction by least-squares matching of image grey values with object surface and object reflection models, e.g. Facets Stereo Vision (FAST Vision) from Wrobel [44,45]. What is in essence new is that matching from image to image is overcome by matching from image to predefined object space models to determine their parameters. This strict area based approach can readily be expanded to include the determination of exterior orientation parameters of all images involved [46]. In contrast to feature based procedures, now, each grey value of a picture representing an imaging ray is integrated into the computation for orientation. No preprocessing for features, above all no search for correspondences is necessary. Features normally exist which are low in number and low in density in the image plane. This situation favors problems with respect to critical surfaces of orientation (Sect. 2.4). Now, with pixel based orientation the redundancy is very high and the risk with critical surfaces will be substantially reduced. Perhaps this new mathematical basis of digital photogrammetry [45] might replace in the future the orientation procedures we favor now.

Table 2.2. Standard orientation tasks of photogrammetry and computer vision.

No.	Combining information from 2D images with [3D] object space* / name of the procedure	Corresponding information			
		points	lines	surfaces	digital picture function
1	**2D-2D** relative orientation (Finsterwalder [2], [3]), recovery of motion [101] (five or six parameters)	•	○	○	○
2	**2D-[3D]** image orientation, space resection [2], [3], camera calibration [12], object pose estimation [34], image-to-database correspondence problem [38] (six parameters)	•	○	○	○
3	**3D-[3D]** absolute orientation [2], [3], 3D similarity transformation (seven or six parameters)	•	○	○	
4	**2D-2D-2D** trinocular stereo vision (non metric)	○	○		
5	**2D-2D-[3D]** (mini)bundle block formation by bundle triangulation ¶ (= bundle block adjustment, Schmid 1958 [51]), binocular stereo vision (metric)	○	○	○	○
6	**2D-3D-[3D]** –				
7	**3D-3D-[3D]** –				
8	**2D-2D-2D-2D** four-camera stereo vision (non metric)	○	○		
9	**2D-2D-2D-[3D]** bundle triangulation¶, trinocular stereo vision (metric)	○	○	○	○
10	**2D-2D-3D-[3D]** –				
11	**2D-3D-3D-[3D]** –				
12	**2D-...-2D** bundle triangulation, image sequence analysis	○	○	○	○
13	**2D-...2D-[3D]** bundle triangulation, image sequence analysis or **2D-2D-2D... ⇒ 3D$_m$-3D$_m$...-[3D]** model block formation by model block adjustment	○ ○	○	○	○

* 3D without [ ] symbolizes a set of points in 3D space not belonging to object space, but e.g. to an individual mobile stereo vision system.

¶ The star of imaging rays of an image is denoted in photogrammetry by a bundle of imaging rays.

• Basic orientation tasks, see Sect. 2.3

○ Existing orientation procedure for which no further information will be given.

Remark: For orientation task no. 1, 4, 8 and 12, there are new solutions based on the projective camera model, with very interesting characteristics with respect to linearity, uniqueness and low computational expense, see [19,29–33,67,68,76]

The orientation tasks no. 1–3 in Table 2.2 are regarded as the basic mathematical approaches for orientation. With these, in a sequential manner practically all configurations of many images, also called "image blocks", can be solved for orientation. When there are sufficient features in one image with given object space coordinates, single image orientation (or space resection) may be used. Exactly this happened historically at first when in the middle of the 19th century the first air photographs from balloons were taken and when photogrammetry came into existence. The classical photogrammetric task of orientation of a stereo pair of images (Table 2.2, no. 5) has traditionally been solved in two steps (Table 2.2, no. 1 and 3): by relative orientation and absolute orientation, first proposed by Finsterwalder [2,3]. He called that task "Hauptaufgabe der Photogrammetrie", due to its importance for 3D reconstruction for mapping. Today, this two-step orientation is still widely applied, a basic tool of photogrammetric stereo workstations.

As mentioned earlier the parameters of image orientation are identical with the six exterior orientation parameters of the perspective camera model (Table 2.1) and can be seen in Fig. 2.5. Relative orientation mathematically is formulated usually with five parameters, describing the relative position and rotation of one image reference frame with respect to another image reference frame, irrespective of a global scale factor (or one translation). The choice of five parameters is not unique. We mention just one choice: one camera is accepted as reference, the other is moved by two translations and three rotations. Relative orientation is usually obtained by reconstructing the intersection of five or more corresponding image rays without any metric information of object space. Fig. 2.6 shows the principle for more than two images if you forget ground control for the moment. Thus, it constitutes a photogrammetric "3D model" of the object surface, simply represented by a set of points. Corresponding points of this 3D model and of the metric 3D object surface are linked by a 3D similarity relationship. It can be established with the use of three or more corresponding points by absolute orientation with seven degrees of freedom (three translations, three rotations and one scale factor). Together, the relative and absolute orientation determine the twelve parameters of the exterior orientation of two images.

There are many applications for which another partitioning of the twelve parameters is imperative, namely into both six parameters, because the aforementioned scale factor has already been included in relative orientation and not in the absolute orientation. We mention just one example. Mobile robots may be equipped with a stereo vision system, calibrated to sense metric 3D scenes in a robot centered coordinate system. To relate a scene to a 3D world coordinate system or to a neighboring scene a six-parameter similarity transformation has to follow [47].

At the end of the discussion on the three basic orientation tasks we have also to address 2D-2D absolute orientation. It is a simple special case of 3D-

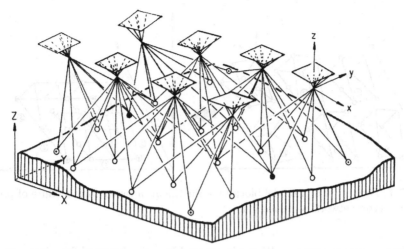

$\odot, \bullet$ control points (reference points) of object space with given coordinates
    X, Y, Z or Z, repectively, arranged on the border of a block
$\circ$ tie point with unknown X, Y, Z.

Fig. 2.6. Bundle triangulation by bundle block adjustment.

3D absolute orientation (not of 2D-2D relative orientation!). Full closed-form equations are given in [34].

Most photogrammetric or computer vision applications cannot be solved with only one or two images. For close-range applications configurations of three or four cameras, see Table 2.2, no. 4–9, have been investigated and successfully used in practice (e.g. the mensuration of industrial parts [48,49], the tracking of tracer particles for velocimetry of fluids [50], and the control of a robot [47]). The orientation of cameras into a configuration with a common 3D coordinate system has sometimes been accomplished simply by independent single camera orientation (space resection) with respect to a common 3D set of given object points [47]. Others, especially photogrammetrists, prefer the more stable orientation of all images simultaneously by bundle triangulation – a least-squares (Gauss–Markov) estimation of all possible intersections of image rays together with integration of given object space coordinates (Schmid [51]), see Fig. 2.6. Bundle triangulation has been thoroughly investigated in photogrammetry in the past. Today, this procedure is operational for thousands of images. Software with sophisticated refinements and statistics is available on the market [52,53]. Even the almost completely automatic bundle solution for digital aerial images has appeared, e.g. offered by the companies LEICA, Switzerland, and ZEISS, Germany [54], [55]. Besides bundle triangulation the quasi two-step alternative of it, block adjustment with independent 3D models, also has been developed (Fig. 2.7). It has proven its capacity over many years of application [52].

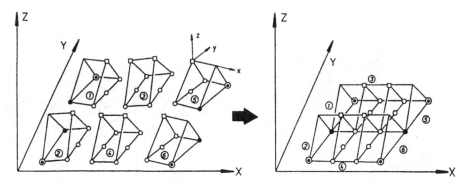

Fig. 2.7. Formation of a model block by adjustment: a two-step procedure of photogrammetric network computation.

Both multi-image or multi-model procedures may be used for corresponding tasks of no. 4–12 in Table 2.2. Also, if we do not interpret the sets of 3D model points in no. 13 as only generated from two perspective images, but alternatively from other 3D sources (e.g. from range images, from surveying instruments), with procedure no. 13 these data as well can be configurated into one 3D world reference frame in a general way and always in a least-squares sense.

For specific applications it may not be possible (e.g. due to time limitation) to first collect the data from all the images or all the models and then to perform off-line a bundle triangulation or model block adjustment, respectively. Research work has therefore been invested into sequential block formation (Mikhail [56], Kratky [57]). The concept of triplets – a sliding minibundle block of three images so to speak – has been developed for quick consistency checks of correspondences and for blunder elimination. More effectiveness is shown by the strict on-line bundle triangulation, a dynamic recursive least-squares estimation, e.g. with Kalman filtering of consecutively incoming image data, see Gruen [58], Gruen and Kersten [59]. All these concepts of photogrammetry remind me very much of the approaches of image sequence analysis, motion analysis [34] or the new work on multiple view tensors [60], [61].

However, the linear Gauss–Markov estimation process for bundle triangulation requires a linearization of the perspective model and for that approximate parameters have to be procured a priori. These values – finally – can be obtained e.g. with the help of the three basic orientation tasks no. 1–3 in Table 2.2. From the long experience of photogrammetry with bundle triangulation two major strategies have come up:

- Images for bundle triangulation often will not be taken, unless a design of the configuration of all images has been made (their number and orientation, etc.) with respect to the desired specifications of the results. The execution of a design at the object can be obtained with photogrammetric cameras normally with an accuracy of better than ±10% for camera positions and ±15° for camera rotations. With these tolerances least-squares bundle triangulation converges iteratively to the correct solution. A considerable enlargement of these tolerances can be reached by using Hamilton's unit quaternions for representation of rotation as Hinsken [66] has shown, see Sect. 2.4.1. According to that strategy the basic orientation tasks are not required.

- In recent years, it is true, there have been many projects for which hand-held medium format cameras or CCD cameras have been used. With these cameras, it is not comfortable to be forced to realize predesigned orientation parameters with each image. There is rather a desire for more flexible bundle triangulation according to simplified rules, but with many more images than before. The handling of so many images and the approximate block formation for entrance into least-squares bundle triangulation should be automated. Since the advent of those cameras, a new strong interest in the three basic orientation procedures, suitable for the above strategy, can be observed in photogrammetry.

At the end of the review on standard orientation tasks we can state the following findings:

- The approaches of photogrammetry and computer vision for orientation converge to practically indistinguishable procedures. Photogrammetry has a longer tradition in theory and practice, starting in the 19th century. In the beginning, procedures were operator-assisted; however a transition to fully general procedures – a prerequisite for automation – can be observed.
- It is not possible to compare here orientation procedures using various criteria. Nevertheless, there seems to be a prominent substance: At the beginning of a procedure as a first step orientation should be solved as simply as possible: no a priori values about the parameters or knowledge about the object is demanded, the solution is unique in full parameter space and with an acceptable accuracy. In essence a topologically correct result with respect to the spatial interrelationships of object points and images is important. Here, the projective camera model can be recommended for all multi-image orientation tasks (Table 2.2), if the number of corresponding points is rather high and their configuration guarantees a stable solution (see Sect. 2.4). In recent times for these tasks excellent computational procedures have been presented (Shashua and Nabab [30], Zeller and Faugeras [67], Niini [68,32]). By transformation of the projective camera model parameters into the physically defined parameters of the perspective camera model continuation with the latter is a good choice. Then in a second step – if still necessary – a least-squares estimation may follow with refinements for calibration, for the stochastic model of the image points (Weng et al. [70], Weng et al. [26], Spetsakis

and Aloimonos [71]), and for integration into a larger configuration of images, see Gruen [53] and Hådem [72], see also Chaps. 4 and 7 in this book.

2.3   Minimum Solutions for Orientation Using Correspondent Points

We are now looking for solutions of the three basic orientation tasks no. 1–3 of Table 2.2 with the perspective camera model involved and using correspondent points at minimum number. From the numerous approaches we select those with a direct solution. A direct solution can be obtained from a linear system or from nonlinear closed-form equations. Some solutions require the computation of eigenvalues and eigenvectors of matrices of low order, 3×3 or 4×4. In that particular case also closed-form computation of the eigenvalues from the characteristic polynomial is possible. Direct solutions of orientation are attractive due to the following properties:

(A) Direct solution provides orientation data without any a priori guess as opposed to an iterative solution.
(B) The number of arithmetic operations is fixed a priori.
(C) The representation of the orientation can be chosen such that its stable solution in full parameter space of the orientation is guaranteed. All critical configurations of the points in object space with which indeterminacy, instability or multiplicity of the solution are linked have to be known, and all multiple solutions have to be produced by the algorithm.
(D) Usually, to combat noise associated with corresponding points the direct approach should procure a least-squares solution from redundant information. Then, the solution should also exist in the limit of nonredundant information.

Properties (A) and (C) are necessary for automation of orientation computation. Property (B) is very important for computation under strict time requirements, e.g. for real-time control of a robot. Finally, for keeping software systems simple a universal approach comprising property (D) is worthwhile. Now, we will look into the literature to see whether there are procedures for all three basic orientation tasks with property (A)–(D). In the survey of literature we have included – as before – early photogrammetric references.

2.3.1   2D-2D Relative Orientation

The relative position up to an unknown scale factor and rotation of an image reference frame in 3D space with respect to another image reference frame (for example two neighboring aerial images in Fig. 2.6) can be reconstructed from the intersections of at least five pairs of corresponding rays from these two images [3]. The intersections are reestablished if pairs of corresponding rays $\boldsymbol{p}_i'$, $\boldsymbol{p}_i''$ and the base vector $\boldsymbol{b}$ are coplanar, thus, if the following triple product of vectors vanishes:

$$\boldsymbol{R}'\boldsymbol{p}_i' \cdot \left(\boldsymbol{b} \times \boldsymbol{R}''\boldsymbol{p}_i''\right) = 0; \quad i = 1,\ldots,n \geq 5 \qquad (2.3)$$

where $\boldsymbol{p}_i'$ is the ray from the first image to a point P_i in the scene, $\boldsymbol{p}_i''$ is the corresponding ray from the second image to the same point P_i in the scene, $\boldsymbol{R}'$ and $\boldsymbol{R}''$ are rotation matrices of the two images and $\boldsymbol{b} = \boldsymbol{x}_0'' - \boldsymbol{x}_0'$ the vector between the projection center $\boldsymbol{x}_0'$ and $\boldsymbol{x}_0''$ of the first and second image, respectively (Fig. 2.5). The base vector $\boldsymbol{b}$ is also called the epipolar axis of the two images. It is geometrically obvious that the epipolar axis is the axis of a pencil of epipolar planes, each defined by a pair of corresponding rays $\boldsymbol{p}_i'$, $\boldsymbol{p}_i''$.

The relative orientation is represented by a choice of five independent parameters. If we choose the first image as reference then $\boldsymbol{R}' = \boldsymbol{I}$ may be taken and we have three parameters for the rotation matrix $\boldsymbol{R}''$ and two for the translations with the base vector $\boldsymbol{b}$ normalized to unit length. (2.3) is the starting relationship for many of the known approaches of relative orientation.

Mathematically the same task arises when a fixed camera takes two images of a moving rigid object from which its motion in 3D space has to be recovered, i.e. the rotation matrix $\boldsymbol{R}''$ and unit translation vector $\boldsymbol{b}$. In this context the task is called recovery of motion. Due to its importance for all kinds of 3D image analysis on one side and its non-trivial mathematical character together with its proneness to errors (measurement, rounding and model errors) on the other side the task of relative orientation attracted much interest in theory and practice.

It is worthwhile to mention here the very first approaches to relative orientation treated in projective geometry by O. Hesse 1863, H. Schröter 1880, G. Hauck 1863 (for details and references see [2,3]), R. Sturm [73] and E. Kruppa 1913 [4]. At that time relative orientation is known as the problem of the "congruent movement" of two stars of rays (usually, in photogrammetry a star of rays is called a bundle of rays) into perspective position yielding polynomial equations. However, in the past the solutions along that line have not been accepted in practice. Only in recent time a strong new, at least theoretical interest came up again to which we refer later.

Now, in the remainder of this section we first regard a special closed-form procedure for plane objects, a so-called four-point algorithm of relative orientation, given by Wunderlich [74] (a hint for clarity: in the name of the algorithm with "point" usually are meant the object points belonging to corresponding pairs of image points or rays). Then we consider procedures for general 3D object surfaces, applicable not only with eight and more points (i.e. the redundant case with least squares estimation) but also with only seven, six or down for the minimum information with only five points. This is just the sensible region of information where relative orientation behaves complicated, but for which since some few years an intelligent unified treatment has been developed. Finally, some remarks are given on the best of iterative procedures.

In case when the object is a plane surface then the relative orientation is determined already by four points, with the condition that no three of them are collinear. Then, the two stars of image rays are in an affine relationship which is determined in 3D space by a specific direct procedure with four points. There are many applications related to plane objects, e.g. indoor scenes or the rectification of images of facades of historical buildings. So it is worthwhile to know a closed-form algorithm of the relative orientation for that specific case. Hofmann-Wellenhof [83] presented 1979 a polynomial solution procedure, however, not fully worked out. In contrast Wunderlich [74] presented such an algorithm which has to be assigned properties (A)–(C), see Sect. 2.3. Its derivation afforded knowledge of intrinsic projective geometry. His basic conception is briefly sketched here.

Wunderlich does not use the coplanarity condition (2.3). He does not compute a pencil of epipolar planes which are congruent to those in the moment of taking the images. Instead, he aims at the relative orientation by direct computation of points which – disregarding a scale factor – are congruent with those on the plane object surface. So, with disposal of the epipolar axis, used in (2.3), the "photogrammetric model", mathematically similar to the plane object, is immediately obtained. Wunderlich starts from the well-known collineation $\boldsymbol{p}_i'' = \boldsymbol{A}\boldsymbol{p}_i'$, $i = 1, \ldots, 4$, that exists between corresponding image points of two images of a plane object. The matrix $\boldsymbol{A}$, here, is a 3×3 matrix with only eight relevant elements. The position vectors $\boldsymbol{p}_i'$, $\boldsymbol{p}_i''$ are related to perspective collineation (Table 2.1, (2.1)), are made homogeneous, but nevertheless are treated as inhomogeneous 3D vectors. This interpretation has been the key step for the successful derivation of the solution. First, matrix $\boldsymbol{A}$ is computed from the given four image coordinate vector pairs, $\boldsymbol{p}_i'$, $\boldsymbol{p}_i''$. Then a triple of orthogonal vectors are sought in both bundles of image rays, which retain its orthogonality under affine transformation with matrix $\boldsymbol{A}$. Of main importance are the eigenvalues and eigenvectors of the matrix $\boldsymbol{A}^T\boldsymbol{A}$, from which straightforward relative orientation finally results. It is represented by two rotation matrices, $\boldsymbol{R}'$ for the first image and $\boldsymbol{R}''$ for the second. Therefore, the oriented bundles of image rays from the two images are still congruent to the original ones. $\boldsymbol{R}'$, $\boldsymbol{R}''$ are defined such that the image planes after rotation are parallel to the object plane. Wunderlich also shows that relative orientation in the case of a plane object with minimum number of four corresponding points has two real solutions, which are both obtained with his algorithm (see Sect. 2.4.3, no. (7)). For details of the derivation reference is made to Wunderlich [74]; computational formulae and one example are given by Kager et al. [75].

We now return to general 3D object surfaces and look at closed-form algorithms at minimum information (= five points) and at situations with six, seven, eight and more points. For better understanding the coplanarity conditions (2.3) are transformed. With

$$\boldsymbol{R}^{'} = \boldsymbol{I}, \quad \boldsymbol{B} = \begin{bmatrix} 0 & -b_3 & b_2 \\ b_3 & 0 & -b_1 \\ -b_2 & b_1 & 0 \end{bmatrix} \quad \text{and} \quad \boldsymbol{B}\boldsymbol{R}^{''} = \boldsymbol{E} = \{e_{ij}\} \qquad (2.4)$$

relationship (2.3) takes on the new form

$$\boldsymbol{p}_i^{'T} \boldsymbol{B}\boldsymbol{R}^{''}\boldsymbol{p}_i^{''} = \boldsymbol{p}_i^{'T} \boldsymbol{E}\boldsymbol{p}_i^{''} = 0 \quad ; \quad i = 1, ..., n \qquad (2.5)$$

where $\boldsymbol{E}$ is denoted as the essential matrix of relative orientation. Since the equations (2.5) are homogeneous it is sufficient to determine the relative size of $\boldsymbol{E}$, relative to one specific element, $e_{33} = 1$ say, or with respect to a norm of $\boldsymbol{E}$, e.g. the Frobenius norm $||\boldsymbol{E}||_F = 1$ Then $n \geq 8$ pairs of corresponding rays $\boldsymbol{p}_i^{'}, \boldsymbol{p}_i^{''}$ are necessary. However, with "$\boldsymbol{E}$", simply determined from system (2.5) alone, its special properties, see (2.4), are not yet enforced. Due to noise and other disturbances this computed "$\boldsymbol{E}$" comes only close to an exact essential matrix $\boldsymbol{E}$. However, its constituents, matrix $\boldsymbol{B}$ and $\boldsymbol{R}^{''}$, have to be derived exactly because they represent at the end the elements of relative orientation (or parameters of motion).

At this point two main lines of closed-form algorithms have emanated in the past. On one line, leading to the eight-point algorithms, matrix "$\boldsymbol{E}$" is computed from system (2.5) with at least $n = 8$ points (in practice $n \gg 8$ to obtain stability). Afterwards, by inversion of (2.4) the matrices $\boldsymbol{B}$ and $\boldsymbol{R}^{''}$ are derived uniquely from "$\boldsymbol{E}$" and in a least-squares sense. On the other line the properties of $\boldsymbol{E}$ are integrated into the solution procedure from the beginning with two consequences: the number of coplanarity equations (2.5) can be reduced down to five point pairs, the minimum information for relative orientation, and $\boldsymbol{E}$ will come out exact (five-point algorithms). From the definition of $\boldsymbol{E}$ being the product of a skew-symmetric matrix $\boldsymbol{B}$ and an orthonormal matrix $\boldsymbol{R}^{''}$ the following important properties follow (for more and proofs see [62], [63], [64], [65]):

- A 3×3 matrix $\boldsymbol{E}$ is an essential matrix if and only if the singular values $\sigma_1 \geq \sigma_2 \geq \sigma_3$ of $\boldsymbol{E}$ satisfy $\sigma_1 = \sigma_2 > 0, \sigma_3 = 0$ (i.e. rank$(\boldsymbol{E}) = 2$ and det$(\boldsymbol{E}) = 0$).
- The elements e_{ij} of an essential matrix satisfy a set of nine homogeneous polynomial equations of degree three:

$$\boldsymbol{E}\boldsymbol{E}^T \boldsymbol{E} = \frac{1}{2} \text{tr} \left(\boldsymbol{E}\boldsymbol{E}^T \right) \boldsymbol{E}. \qquad (2.6)$$

Therefore, the task of relative orientation is equivalent to the problem of finding all elements e_{ij} of $\boldsymbol{E}$ from the set of $n = 5$ linear equations (2.5) together with the common zero of the polynomial equations (2.6) and with $\|\boldsymbol{E}\|_F = 1$.

Some remarks are added about these most important lines of algorithms for relative orientation.

There are very many direct linear procedures of eight-point relative orientation, some of them were presented long time ago, e.g. by von Sanden 1908 [24]. Most of them have in common relationship (2.5), but for further derivations, e.g. the computation of $\boldsymbol{R}''$ and $\boldsymbol{B}$ from $\boldsymbol{E}$, treatment of noise in the image points, they move in different directions. Some rely on algebraic projective geometry (projective image correlation, reconstruction of epipoles, etc.), see von Sanden [24], Thomson [23] and Brandstätter [19–21] for details and references in photogrammetry; in computer vision there are e.g. Maybank and Faugeras [29], [27], [28], [63], Hartley [76]. The other approaches rely more on algebraic geometry, e.g. Rinner [22], Stefanovic [77], also van den Hout and Stefanovic [78]. Apparently Stefanovic in 1973 was the first to investigate some properties of the polynomial system (2.6) and to use it together with the equations (2.5) to yield an optimal and exact matrix $\boldsymbol{E}$, however, iteratively. This procedure is still an 8-point algorithm: He first solves the linear system (2.5) for an approximate E, then enters into a least-squares adjustment with linearized versions of the systems (2.5) and (2.6) and with the image coordinates of the corresponding points as stochastic variables. Within very few iterations the optimal and exact matrix $\boldsymbol{E}$ is obtained. In computer vision Longuet-Higgins [25] seems to be very popular with his 8-point algorithm from 1981, there pioneering the further work on relative orientation. One major subject is the low stability of the system (2.5), being very susceptible to all kinds of perturbations, which gave rise to continuous improvements (e.g. use of orthogonalization algorithms, eigenvectors, eigenvalues, singular value decomposition instead of Gauß elimination [79]) presented by Weng et al. [70], Horn [80], see also [28], [65]. Besides differences in functional relationships there are also differences in strictness of treating noise, see criticism and progress given by Spetsakis and Aloimonos [71], also Kanatani [81]. Torr and Murray [82] compare various stochastic models and robust estimation procedures. Realistic weighting with respect to the proper stochastic noise source, which are the image coordinates of corresponding features, has been presented earlier by van den Hout and Stefanovic [78], Horn [80] and Brandstätter [20]. Looking back at all these proposals I have the impression that with these noise refinements the simplicity and easy handling of the basic linear approach (2.5) gets lost.

All these eight-point procedures of relative orientation are characterized by a rather grave drawback. Due to the increase of degrees of freedom from five to eight parameters when using relationship (2.5) instead of (2.3) a corresponding affine deformation of the bundle $\boldsymbol{p}_i''$ of the second image is admitted. Consequently a plane object surface is already a critical surface for orientation (see von Sanden [24]) and additionally all surfaces of second degree if they comprise also the two centers of perspective; for proofs see Krames [5], p. 334. In contrast, with five-point relative orientation no affine deformation of the two bundles of rays is allowed and with points on a plane object surface, generally a stable orientation is feasible. More on stability and critical configurations follows in Sect. 2.4.3.

The second line of algorithms for relative orientation makes possible the solution with the minimum information of five point pairs only. It is owing to O. Faugeras and his group [110], [28], [67], [63] that Kruppa's early work [4] in projective geometry on relative orientation from 1908 has been revisited and made known to the computer vision community giving rise to excellent new investigations. Faugeras reexamined Kruppa's polynomial representation of 10th degree, clarified that there are up to ten solutions (only few of them can be real, see Sect. 2.4.3) and conveyed numerical tests. Not very surprising the results revealed a considerable susceptibility to small perturbations in the corresponding point pairs. The comment of O. Faugeras himself on this five-point algorithm reads [28, page 278]: "It should be considered more as a proof of the complexity of the problem than as a way of solving it." Probably the latest developments are presented by J. Philip [64] in 1996. He describes in detail a set of closed-form solution procedures covering not only the case for $n = 5$ point pairs, but also the cases for 6, 7 and $n \geq 8$ point pairs, thus filling the algorithmic gap between $n = 8$ and $n = 5$. Philip's derivations are inspired by the previous work of Hofmann-Wellenhof [83] on closed-form relative orientation, however, Philip completes it with respect to n down to $n = 5$ and improves its algorithmic performance (e.g. detection of critical configurations, etc.). Both, Hofmann-Wellenhof and Philip, apply a general approach of Killian and Meissl [86] for solving geometrical problems that can be expressed as a system of polynomial equations. Its major characteristic is that redundancy must be exactly one to make possible a Gauss-Jordan like elimination process down to a tractable system. This approach had already been successfully used for the 2D-3D image orientation [87]. Besides the attractive solution mechanism, relative orientation with $5 + 1$ points in general position (Sect. 2.4.3, no. (4)) leads to a linear system and to a unique solution, whereas with five points there may be up to two or three real solutions (Sect. 2.4.3). Philip performed a large number of computational tests ([64], [65]) with the main findings that with $n = 6$ numerically correct solutions have been reached, and with $n = 5$ the results are not reliable (even exceeding the number of real solutions although limited by theory). Thus the above given comment of O. Faugeras is confirmed.

Finally, for comparison reasons, the state-of-the-art of iterative relative orientation for five and more points is mentioned briefly. Algorithms of that kind have been applied with success (and with care!) in photogrammetric production since the 1920s [52]. Usually a Taylor series expansion of the refined perspective camera model (2.2), Table 1, for the two images with respect to five parameters is taken, wherein the coordinates of the corresponding image points p_i' and p_i'' are treated as stochastic variables. As parameters often three rotation angles and two components of the base vector b are chosen. The procurement of approximate parameters for the series expansion with aerial photography (Fig. 2.4 and 2.6) is no problem at all, in contrast to close-range applications of photogrammetry.

The two approaches, of Hinsken [66] (his full equations are also in [34]) and of Horn [80], are apparently the furthest developments. Hinsken adopts the parameterization of rotation given by Pope [88] (Hamilton's unit quaternions plus very simple yet strict update formulae for iterative computation, no singularities in full parameter space) and transfers it to the standard orientation tasks no. 1–4, 8 and 12 of Table 2.2. His procedure for relative orientation is iterative; however, the start values may differ from corresponding correct angles to about $50°$. For many applications this amount is fully sufficient.

Horn's paper conveys a very coherent insight into the variety of approaches of relative orientation and their problems, seen from various aspects. His approach – also based on quaternions, also iterative – has to be supplied as well with sufficiently close initial approximations for the parameters. However, he proposes a systematic sampling of parameter space, controlled by appropriate error functions, to defeat the dependence on sufficient initial parameters. So, both methods are iterative in nature, but nevertheless show attractive properties for automation.

At the end of this section the reader might ask himself whether the relative orientation of two images might be useful at all for operational image analysis. It is true at minimum information the results definitely are not reliable. However, photogrammetry applies relative orientation routinely as one of its core tools with the operator controlled photogrammetric stereoscopic workstations. This has been demonstrated successfully since the 1920s and is still ongoing today. Of course, being aware of its instability some rules of operation for relative orientation are essential: introduce image coordinates corrections from a camera calibration (such as lens distortion), avoid instability from the beginning through too small intersection angles of corresponding rays by admitting only reasonable base-to-distance ratios of about 1 to 0.1 (i.e. the length of the base vector separating the two images in relation to the mean camera distance to the scene), and last not least in order to yield accuracy and reliability realize a very high redundancy, whether by visual stereoscopic inspections and corrections or by the number of corresponding point pairs. The last item has become the strategy of digital photogrammetry with its new and automatic procedures of relative orientation, where the redundancy amounts to fifty or even more point pairs [54].

Let us summarize the findings about 2D-2D relative orientation: There exist closed-form orientation procedures for the minimum information of $n = 5$ point pairs and for the redundant case $n > 5$. The procedures show properties (A)–(D), see Sect. 2.3. At minimum information the solutions may be

multiple and are not reliable. Therefore, for practical work a considerable
redundancy of point pairs is recommended.

2.3.2   2D-3D Image Orientation (Space Resection)

Given a set of corresponding points P_i', P_i with its space vectors $\boldsymbol{p}_i'$, $\boldsymbol{p}_i$ ($i =
1, \ldots, n \geq 3$) from an image and 3D object, respectively (see the perspective
camera model in Table 2.1 and Fig. 2.8), the process of image orientation
is defined as the computation of its exterior orientation parameters, i.e. the
position X_0, Y_0, Z_0 of its center of perspective and the rotation of the image
reference frame, represented e.g. by appropriate rotation angles ω, φ, κ or
by quaternions a, b, c, d. As can be seen in equation (2.2) in Table 2.1 the
minimum solution should be obtained with $n = 3$ noncollinear points. Indeed,
a closed-form algorithm for minimum information exists with properties (A)–
(C), Sect. 2.3. With $n = 3$ points there are in general four solutions of
orientation (Sect. 2.4.4).

The general perspective camera model simplifies to a 2D-2D similarity
transformation with four parameters (one rotation, two translations and one
scale factor) in the case where all object points are positioned on one plane
and where image plane and object plane are parallel. Image orientation in
that specific case is obtainable with $n = 2$ corresponding points. This task
represents a simple standard problem of photogrammetry and image processing. Algorithms may be found in corresponding textbooks, see [52], [34],
where also formulae for robust parameter estimation are given. By the way,
the same similarity relationship may result as a special case of 3D-3D absolute
orientation; see the following subsection.

2D-3D image orientation was historically the first orientation procedure
to be applied in photogrammetry, namely since the middle of the 19th century. For its most difficult part investigations and solutions already existed,
originating from the work of Lagrange [1] (iterative solution of polynomials)
and G. Monge 1798 [84] (graphical solution). The first closed-form algebraic
solution apparently is from Grunert 1841 [85]. Since that times, due to its
importance for applications of photogrammetry and for the same reason in
computer vision in recent time, publications on 2D-3D image orientation may
be regarded as a never ending story.

I guess now there are altogether more than one hundred more or less different approaches that have been published if we consider approaches based
on projective and perspective camera models and on mixtures of both. As I
said before we will concentrate here only on solutions with the perspective
camera model. Readers interested in closed-form solutions with the projective
camera model are refered to Abdel-Aziz and Karara [11] (six-point eleven parameter solution), Shih and Faig [13] (extended eleven parameter approach
for 3D object surfaces with more than six points and four-point eight parameter solution specialized to 2D object surfaces), Liu et al. [90] (three-point
and six-point solution), and Weng et al. [26] (six-point solution).

30   Bernhard P. Wrobel

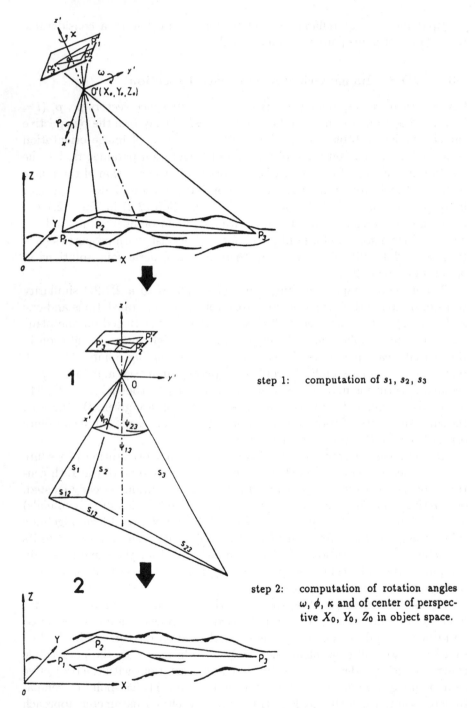

step 1:   computation of s_1, s_2, s_3

step 2:   computation of rotation angles ω, ϕ, κ and of center of perspective X_0, Y_0, Z_0 in object space.

Fig. 2.8. Image orientation: a three-point closed-form algorithm in two steps.

As already mentioned a four-point solution (minimum points + 1) for image orientation with the perspective model was presented by Killian [87], [86], also with simpler formulae, specialized for horizontally plane objects. What makes an approach with four points so interesting is that orientation will already have a unique solution (not up to four solutions as with three points), if all four points are positioned on one plane, even if they are on the so-called "dangerous cylinder", see Sect. 2.4.4 for details. Killian's approach of image orientation is integrated as a basic tool into photogrammetric workstations of the LEICA company, Switzerland (private communication of Dr. Bormann).

Now, let us have a closer look at one of the three-point closed-form solutions of image orientation. You may find such procedures traditionally in every textbook on photogrammetry [52] or computer vision [34], however I recommend the following recent papers of Lohse et al. [91], [?] and Haralick et al. [89], where complete computational equations and derivations are given. The solution of (2.2) in Table 2.1 is obtained in two steps. Figure 2.8 describes the geometry of the approach. The space vectors p'_i, p_i of the three given corresponding points P'_i, P_i define a tetrahedron, with its vertex identical with the center of perspective $O'(X_0, Y_0, Z_0)$, and its base triangle formed by the object space points P_1, P_2, P_3. From these data with simple trigonometric relationships the triangle side lengths s_{12}, s_{13}, s_{23} and face angles ψ_{12}, ψ_{13}, ψ_{23} can be derived easily. The proper mathematical problem of space resection is the derivation of the side lengths s_1, s_2, s_3 of the base triangle of the tetrahedron. This nontrivial problem is solved in step 1 of space resection, whereas the rest can be realized immediately in step 2. For step 1 the law of cosines in each of the three triangles of the tetrahedron gives the following set of equations:

$$
\begin{aligned}
s_{12}^2 &= s_1^2 + s_2^2 - 2s_1 s_2 \cos \psi_{12} \\
s_{23}^2 &= s_2^2 + s_3^2 - 2s_2 s_3 \cos \psi_{23} \\
s_{31}^2 &= s_3^2 + s_1^2 - 2s_3 s_1 \cos \psi_{31} .
\end{aligned}
\tag{2.7}
$$

From these polynomial equations of second degree in three unknowns s_1, s_2, s_3 the solution has to be derived. It is known from algebra that this system of three unknowns can have no more solutions than the product of their respective degrees, i.e. $2 \cdot 2 \cdot 2 = 8$. Since every term in (2.7) is either a constant or of second degree, therefore, for every real positive solution there exists a geometrically isomorphic negative solution (see Sect. 2.4.4). So, in general four real solutions are possible. The correct solution can be found with some additional information, e.g. with a fourth given point. Haralick et al. [89] have investigated the algorithms of Grunert 1841 [85], Finsterwalder and Scheufele [3], Merrit 1949 [92], Fischler and Bolles [38], Linnainmaa et al., and Grafarend et al. [91,93] and compared them with respect to numerical accuracy and stability. Haralick et al. ascertain the clear superiority of the algorithm of Finsterwalder.

After the solution of step 1 the scale factor λ_i of each image vector $\boldsymbol{p}'_i$ in the equation of perspective collineation (Table 2.1) now can be computed from the relationship

$$\frac{1}{\lambda_i} = +\frac{\sqrt{\boldsymbol{p}_i'^{\mathrm{T}} \boldsymbol{p}'_i}}{s_i} \quad i = 1, 2, 3 . \tag{2.8}$$

Next the new scaled image point vectors $\boldsymbol{q}'_i$ are determined from

$$\boldsymbol{q}'_i = \frac{1}{\lambda_i} \boldsymbol{p}'_i = \left(X'_i, Y'_i, Z'_i\right)^{\mathrm{T}} \quad i = 1, 2, 3 . \tag{2.9}$$

The vectors $\boldsymbol{q}'_i$ have the same length as the edges s_i of the tetrahedron, but are still defined with respect to the image reference frame. With (2.9) the equation of perspective collineation now reads:

$$\boldsymbol{q}'_i = \boldsymbol{R}\boldsymbol{p}_i + \boldsymbol{t} ; \quad i = 1, 2, 3 . \tag{2.10}$$

So, in step 2 of image orientation we have to solve a six-parameter 3D similarity transformation (absolute orientation) from three corresponding points. For that according to Schut [95,91] the following very elegant linear solution is attractive. The rotation matrix $\boldsymbol{R}$ is parameterized in normalized Hamilton quaternions a, b, c, d, with one parameter chosen equal to unity, e.g. $d = 1$; also denoted as Cayley parametrization [103]. We obtain $\boldsymbol{R}$ as an analytical function of a skew-symmetric matrix $\boldsymbol{S}$ with three independent parameters of rotation a, b, c:

$$\boldsymbol{R} = (\boldsymbol{I} - \boldsymbol{S})^{-1} (\boldsymbol{I} + \boldsymbol{S}) \quad ; \quad \boldsymbol{S} = \begin{bmatrix} 0 & -c & +b \\ c & 0 & -a \\ -b & a & 0 \end{bmatrix} \tag{2.11}$$

$$\boldsymbol{R} = \frac{1}{1 + a^2 + b^2 + c^2} \begin{bmatrix} 1 + a^2 - b^2 - c^2 & 2(ab - c) & 2(ac + b) \\ 2(ab + c) & 1 - a^2 + b^2 - c^2 & 2(bc - a) \\ 2(ac - b) & 2(bc + a) & 1 - a^2 - b^2 + c^2 \end{bmatrix} . \tag{2.12}$$

Due to the decomposition (2.11) of the rotation matrix $\boldsymbol{R}$, (2.10) yields a linear system:

$$(\boldsymbol{I} - \boldsymbol{S}) \boldsymbol{q}'_i = (\boldsymbol{I} + \boldsymbol{S}) \boldsymbol{p}_i + (\boldsymbol{I} - \boldsymbol{S}) \boldsymbol{t} \quad ; \quad i = 1, 2, 3$$

or

$$-\boldsymbol{S} \left(\boldsymbol{q}'_i + \boldsymbol{p}_i\right) + \boldsymbol{u} = \boldsymbol{p}_i - \boldsymbol{q}'_i ; \quad i = 1, 2, 3 \tag{2.13}$$

where $\boldsymbol{u}$ is defined by the change of the parameters:

$$\boldsymbol{u} := -(\boldsymbol{I} - \boldsymbol{S})\boldsymbol{t} = (u, v, w)^{\mathrm{T}} . \tag{2.14}$$

After expanding (2.13) and ordering according to the six unknown parameters a, b, c and u, v, w we obtain:

$$
\begin{bmatrix}
0 & -Z_i' - Z_i & Y_i' + Y_i \\
Z_i' + Z_i & 0 & -X_i' - X_i \\
-Y_i' - Y_i & X_i' + X_i & 0
\end{bmatrix}
\cdot
\begin{bmatrix} a \\ b \\ c \end{bmatrix}
+
\begin{bmatrix}
1 & 0 & 0 \\
0 & 1 & 0 \\
0 & 0 & 1
\end{bmatrix}
\cdot
\begin{bmatrix} u \\ v \\ w \end{bmatrix}
=
\begin{bmatrix} X_i - X_i' \\ Y_i - Y_i' \\ Z_i - Z_i' \end{bmatrix}
$$
(2.15)
$$ i = 1, 2, 3. $$

With three given points there are altogether nine equations; however, an adequate choice of only six equations will form an independent set of equations, which gives the correct solution, e.g. with three equations from $i = 1$, the first two from $i = 2$ and the last one from $i = 3$:

$$
\begin{bmatrix}
0 & -Z_1' - Z_1 & Y_1' + Y_1 & 1 & 0 & 0 \\
Z_1' + Z_1 & 0 & -X_1' - X_1 & 0 & 1 & 0 \\
-Y_1' - Y_1 & X_1' + X_1 & 0 & 0 & 0 & 1 \\
0 & -Z_2' - Z_2 & Y_2' + Y_2 & 1 & 0 & 0 \\
Z_2' + Z_2 & 0 & -X_2' - X_2 & 0 & 1 & 0 \\
-Y_3' - Y_3 & X_3' + X_3 & 0 & 0 & 0 & 1
\end{bmatrix}
\cdot
\begin{bmatrix} a \\ b \\ c \\ u \\ v \\ w \end{bmatrix}
=
\begin{bmatrix}
X_1 - X_1' \\ Y_1 - Y_1' \\ Z_1 - Z_1' \\ X_2 - X_2' \\ Y_2 - Y_2' \\ Z_3 - Z_3'
\end{bmatrix}.
$$
(2.16)

The geometrical interpretation of this choice is obvious. For better appreciation of taking a reduced choice of equations, remember that the 3D similarity transformation (2.10) is just one step of image orientation which altogether with only three points is not overdetermined.

After the solution of system (2.16) the rotation parameters a, b, c are inserted in (2.11), (2.12) to yield S and R. The matrix S and vector u are introduced in (2.14) for final computation of t, the position vector of the center of perspectivity. Note that the computations of step 2 have to be carried out just as many times as there are real solutions from step 1. That completes the closed-form solution of 2D-3D image orientation at minimum information.

The one-to-four solutions may not exist if the Cayley parameterization (2.12) is used for rotations of 180° (about a fixed axis [103]) or close to it. However, in Sect. 2.3.3 procedures for 3D-3D absolute orientation of step 2 are given without that restriction. So, here we may summarize that for 2D-3D image orientation direct minimum solution procedures exist with properties (A)–(C), see Sect. 2.3.

2.3.3 3D-3D Absolute Orientation

Given a photogrammetric 3D model, represented by a set of points with its 3D space vectors p_i' in an image reference frame, absolute orientation has to translate the model by a translation vector t, to rotate it by a rotation matrix R, and to scale it by the scale factor λ in order to optimally fit the model

into object space, which is represented by a corresponding set of points with
3D space vectors $p_i, i = 1, \ldots n \geq 3$:

$$p_i = \lambda R p_i' + t \qquad (2.17)$$

where p_i, p_i' are vectors of corresponding points from the 3D object space
coordinate system and 3D model reference system, resp.;
$R, R^T R = I$ is the 3×3 orthonormal rotation matrix; $t = (t_x, t_y, t_z)^T$ is the
translation vector, and λ is a scale factor.

Relationship (2.17) is called a 3D similarity transformation with seven
parameters (three rotations for R, three translations, one scale factor). It
may be simplified with $\lambda = 1$ to six parameters, if rigid body motion has to be
recovered. In the context of 3D-3D absolute orientation it may be interesting
to mention also the very special case of 2D-2D absolute orientation. For
example, if the image plane of a camera is mounted strictly parallel to an
inspection table for plane objects, the camera model (Table 2.1) degenerates
to the 2D-version of relationship (2.17) with only four parameters. For closed-
form solutions, see [34], p. 160. We now outline direct least-squares solutions
for the 3D-3D absolute orientation (2.17), but which are also valid for the
minimum number of $n = 3$ points in both systems. Note with three points
there is still a redundancy of two coordinates so that the three-point least-
squares solution still makes sense! Nevertheless, full formulae in the case of
only seven coordinates can be found in Rinner and Burkhardt [96], p. 409.

In photogrammetry the first direct solution was given by Finsterwalder
[3]. He applied elements from mathematical physics of his time, which are
still used today: quaternions for rotation, centroids of the sets of points
for decoupling rotations and scale from the translations. Apparently no further investigation followed by him. The interest in direct numerical solutions
started again in the 1950s since the invention of digital electronic computers.
A method due to Thompson 1959 [94], with improvements from Schut [95]
presents exact linear equations for computation of the rotational elements
of absolute orientation (matrix in Cayley parameterization (2.12)). Tienstra
[97] solved the complete absolute orientation by a direct procedure. He was
obviously the first to transform the least-squares relationships of absolute orientation into an equivalent eigenvalue problem. However, with his approach
nineteen eigenvalues (corresponding to all nine elements of R, six Lagrange
multipliers of orthogonality of R, three translations and one scale factor)
have to be determined, a grave disadvantage! The complete closed-form and
effective least-squares solution of absolute orientation was finally presented
by Sansó 1973 [98]. He makes use of Hamilton's unit quaternions and suc-
ceeds in transforming the problem into a 4×4 matrix, the elements of which
are combinations of sums of products of the corresponding coordinates of
the points. The least eigenvalue and its normalized eigenvector have to be
computed from that matrix. Sansó shows that his solution is unique under a
mild condition (see Sect. 2.4.1) with $\det |R| = +1$ and that it also holds if

all points P_i are coplanar, a special yet possible case of absolute orientation. Later, the same approach of Sansó was given by Horn [99,121], Faugeras and Hebert [108] and others. Another very nice least-squares approach to absolute orientation has been developed through a singular value decomposition of a certain matrix, given by Arun et al. 1987 [100] and Haralick and Shapiro, see [34]. However, it is not guaranteed that R is a proper orthogonal matrix: it might have determinant -1. A comparative study of these methods and some others for rigid body motion recovery may be found in [101]. Now, if there is only the minimum number of three corresponding points, they are certainly all in one plane. Anyway, this simplifies the above mentioned procedures and has been exploited to very effective equations, still preserving the least-squares optimality of the result [99]. Thus, for the 3D-3D absolute orientation task, indeed, there exist effective direct procedures for minimum information and with properties (A)–(D), defined in Sect. 2.3.

2.4   Uniqueness Conditions of Basic Orientation Tasks

As with many other geometrical tasks solved indirectly the previously discussed three basic orientation procedures may yield multiple solutions or unstable solutions or even indeterminate solutions. Problems of that kind arise most likely with a minimum number of correspondent feature points from which an orientation has to be computed. They are not excluded if there are very many corresponding points. Three sources are associated with that kind of phenomenon:

(1) an improper representation of rotation as part of the collineation relationship between image space and object space,
(2) a so-called critical or dangerous configuration of points and of center(s) of perspectivity in object space, and
(3) strong noise of the correspondent points in object space and/or image space.

For automated application of computer stereo vision we have to know completely the strict conditions of uniqueness of orientation. 3D-3D absolute orientation (3D similarity transformation) does not deserve particular attention. It can be regarded as the trivial case as long as a suitable representation of rotation has been chosen (Sect. 2.4.1), and all correspondent points are distributed all over the scene and not on one line. There is another idea that can be clarified at once, namely the impact of the size of rotation angles of an image on the uniqueness conditions of orientation. Actually, there is no direct influence. Of course, the size of the angles is part of the configuration design and have to provide an effective exploitation of the whole format of an image. However, the position of their centers of perspective are really decisive as will come out soon. To deal with these problems in a comprehensive manner in this chapter we first discuss suitable representations of rotation, then explain geometrically the term critical configuration and, finally, we present for 2D-2D relative orientation and 2D-3D image orientation the findings from the literature on multiplicity of solutions, on critical configurations and collect the conditions of uniqueness as recommendations on how to escape these problems. The instability of orientation seen from a statistical point of view will be dealt with by Förstner in Chap. 3.

2.4.1 Representation of Rotation

Three angles for representation of a proper 3×3 orthogonal rotation matrix R (i.e. with its determinant equal to $+1$) are very popular in photogrammetry and computer vision [102,107,52,47]. Also the Cayley parameterization of R, see (2.12), with its three algebraic rotation parameters has been applied due to its low computational cost. However, these two and all the other representations with only three parameters are associated with some unfavorable properties, namely: (a) There are always two parameter sets representing the same rotation matrix R. (b) There are singularities of the differential rotational relationships between object space and image space in some regions of the parameter space. In these critical regions indeterminacy, at least high instability of the computational equations, will arise, as has been shown numerically by Wrobel and Klemm [107] with respect to the spectral condition number and algebraic correlation of the equations. (c) With just one representation it is not possible to represent rotations in a global way in full parameter space; there are at least four representations (= an atlas of at least four charts) required. So, it is inevitable to cast a glance on these properties.

The theoretical foundation of rotations in 3D has been investigated over a long period of time with Euler's theorem 1776 as an early milestone showing that the group of rotations in 3D space is itself a 3-dimensional manifold. Full enlightenment has been found not before the 20th century when the topology of the rotation group SO(3), i.e. the set of all orthonormal rotation matrices in 3D with determinant $+1$, became understood. For discussions and proofs see Stuelpnagel [103], Altmann [104], Kanatani [105], Faugeras [28], also Kühnel and Grafarend [106] with their very interesting and detailed presentation. Now, in the areas of application the following actions with respect to property (a) to (c) can be observed.

To deal with property (a) is very simple. For example: in terrestrial photogrammetry matrix R is preferably represented by the angles φ, ω and κ referred to the axes (really or virtually) of a cardanic suspension of the camera used (in the past called a photo theodolite). Then the second set besides φ, ω, κ reads: $\varphi + \pi, \omega - \pi, \kappa + \pi$. However, by construction of the suspension (or by a corresponding interpretation of the camera movement) the angle ω is restricted to $-\pi/2 \leq \omega \leq \pi/2$, so that one single set of angles exists in practice that satisfy the rotation matrix R.

To overcome the problems encountered with property (b) and (c) different procedures have been applied in practice:

– If it is a priori known that certain 3D rotations will never occur then a representation with three Euler angles (in one of its variants) has to be chosen such that its regions of singularities are well within that section of not occurring rotations [103].

In photogrammetry often two instead of one representation of the rotation matrix $\boldsymbol{R}$, both based on rotation angles, have been applied in combination, however with their critical regions of rotation parameters disjoint [107].

For example: $\boldsymbol{R}$ may be decomposed into the product of three simple rotation matrices each representing a rotation by an angle ω, φ, κ about one of the image coordinate axes x, y, z, respectively, e.g. $\boldsymbol{R} = \boldsymbol{R}_1 = \boldsymbol{R}(\omega; x) \cdot \boldsymbol{R}(\varphi; y) \cdot \boldsymbol{R}(\kappa; z)$, where here rotations are defined in the sequence ω first, then φ and κ with respect to the x axes first, then rotated about y, z accordingly. Just by A change of the sequence of rotations a new representation $\boldsymbol{R}_2$ for the same matrix $\boldsymbol{R} = \boldsymbol{R}_2 = \boldsymbol{R}'(\varphi'; y) \cdot \boldsymbol{R}'(\omega'; x) \cdot \boldsymbol{R}'(\kappa'; z)$ is obtainable with regions of instabilities not overlapping with those of the first representation $\boldsymbol{R}_1$. So, during the computation of $\boldsymbol{R}$ a decision has to be taken depending on the size of the rotation parameters for the appropriate representation which finally should yield a stable solution. However, in view of the theory [106] with both actions no complete representations of rotations (there should be at least four representations!) are established.

– As mentioned earlier in applications of photogrammetry often rather good approximations of rotation angles can be made available, so that a priori an approximation $\boldsymbol{R}_o$ of $\boldsymbol{R}$ can be computed from almost any representation. Therefore, $\boldsymbol{R}$ may be decomposed into the double rotation matrix $\boldsymbol{R}_o \mathrm{d}\boldsymbol{R} = \boldsymbol{R}$, where $\mathrm{d}\boldsymbol{R}$ is a rotation matrix restricted to small rotation angles only. For small rotations there are many very stable representations. So, practically with just one representation of $\mathrm{d}\boldsymbol{R}$ the problem can be solved as well, see [107].

– The representation of rotations by Hamilton's unit quaternions – a representation of rotation in 3D space by four parameters – offers outstanding properties: unrestricted representation of differential rotations in full parameter space and differentiation of rotations is particularly easy (see also [88], [66]). The same rotation matrix $\boldsymbol{R}$ is represented by two quaternions $+\boldsymbol{q}$ and also by $-\boldsymbol{q}$ (for a global 1-1 representation of rotations a minimum number of five parameters is required). An explicit least-squares solution for $\boldsymbol{q}$ has also been found – as already mentioned earlier [98], [108]. From a computed unit quaternion $\boldsymbol{q} = (d, a, b, c)^{\mathrm{T}}$ with $d^2 + a^2 + b^2 + c^2 = 1$ the rotation matrix $\boldsymbol{R}$ follows immediately:

$$\boldsymbol{R} = \begin{bmatrix} d^2 + a^2 - b^2 - c^2 & 2(ab - dc) & 2(ac + db) \\ 2(ab + dc) & d^2 - a^2 + b^2 - c^2 & 2(bc - da) \\ 2(ac - db) & 2(bc + da) & d^2 - a^2 - b^2 + c^2 \end{bmatrix}. \quad (2.18)$$

Also, following Rodrigues (in [104]) the algebraic components of q can be given a very nice geometrical interpretation. Due to the equivalence of a rotation matrix in 3D space with only one rotation axis, given by its unit direction vector $n = (n_x, n_y, n_z)^{\mathrm{T}}$ and by one angle θ, say, we obtain:

$$q = (\cos(\theta/2), n_x \sin(\theta/2), n_y \sin(\theta/2), n_z \sin(\theta/2))^{\mathrm{T}}. \qquad (2.19)$$

If the rotation angle θ, according to the circumstances of the application, can be restricted to $0 \leq \theta < \pi$, then we can impose the condition that the first element d of a quaternion q must be positive. Thus the mapping between rotation and quaternion is unique under this constraint. Without that restriction another three representations are necessary in order to complete the global representation of rotations by normalized quaternions (= an atlas of exactly four charts). The first chart is given with (2.19) and is characterized by $d \neq 0$ and $\theta \neq \pi$. Because of $d \neq 0$ the normalization of q is allowed by dividing every element of (2.19) by d yielding $1, (a/d), (b/d), (c/d)$, and the rotation matrix R in Cayley parameterization (2.12). The other three charts are formed similarly with $d = 0$ and $a \neq 0, b \neq 0, c \neq 0$, respectively. They are capable of rotations $\theta = \pi$, but exclude rotations with $\theta = 0$ and rotations with an axis parallel to the yz-, zx-, and xy-planes, respectively [28].

– Similar advantages as with quaternions have been obtained by direct computation of the rotation matrix R from a singular value decomposition of a certain matrix for 2D-3D orientation [100,34]; see also the representation with a rotation vector, with its length equal to the rotation angle and its direction identical with the axis of rotation [47].

2.4.2 The Terms Critical Configurations of First and Second Kind: a Geometrical Interpretation by Two Examples

The representation of rotations in 3D space as a possible source of instability for image orientation can be avoided once for all by the choice of an appropriate model. However, there still exists the already mentioned second source of instability, namely that of critical or dangerous configurations: i.e. specific relative positions of points and of center(s) of perspectivity in 3D space used for orientation may be a catastrophe for a desired unique and stable solution. In preparation of detailed information on critical configurations in the next section let us have a quick look at two very simple examples in order to explain the important terms, see Fig. 2.9 and Fig. 2.10. Precise mathematical definitions can be found in the references.

In Fig. 2.9 a new point P has to be determined by plane resection. There are three circles c, c_α and c_β. Circle c is defined by three fixed points P_1, P_2, P_3 and circles c_α, c_β are defined by two of the three points and by an angle α, β, respectively. In Fig. 2.9a intersection of the circles c_α, c_β provides a unique

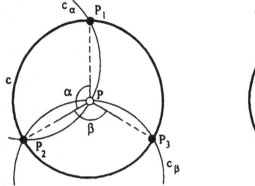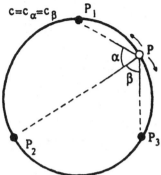

new point P on circle c is undetermined
⇒ **dangerous circle**

a: Stable solution   b: Indeterminate solution

Example: Determination of point P by plane resection with two measured angles α and β

●   given points P_1, P_2, P_3 defining circle c

c_α, c_β circles defined by α, P_1, P_2; β, P_2, P_3, respectively

O   new point P

Fig. 2.9. Critical configuration of the first kind: indeterminacy.

and stable position of the new point P. In Fig. 2.9b, however, the new point P is lying on circle c, so that now all three circles are completely identical $c = c_\alpha = c_\beta$, and, therefore, curvatures and tangents are identical, showing in the direction of indeterminacy. In this particular position of P both α, β are angles at the circumference and therefore they do not change when P moves along that circle. Consequently the computations for the position of P can only produce the result that P actually should be somewhere on that circle. Clearly, this is not a determinate solution, it is useless for practical purposes, and therefore this configuration is called a critical or dangerous configuration of the first kind. In this particular example the locus of indeterminate positions of P is represented geometrically by a circle (= "dangerous circle" of plane resection).

In Fig. 2.10a a new point P again has to be determined by the intersection of two circles c_1, c_2, here defined by measurement of two distances. The critical configuration in this case, Fig. 2.10b, is characterized geometrically by the condition that the new point P is no longer determined by the intersection, but by first order osculation of the two circles c_1 and c_2, i.e. having a common tangent, but no common curvature. Further, a very interesting general feature is that the direction of the common tangent (= direction of instability) shows off the critical line, and allows for a one-parameter differential movement of

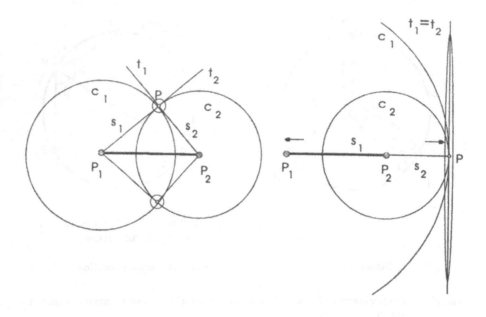

a: Stable solution

two discrete solutions:
numerically well-conditioned,
error ellipse is small and well shaped,
P is well determined

b: Unstable solution

two solutions coincide, also $t_1=t_2$:
numerically ill-conditioned or even
singular,
error ellipse is elongated along $t_1=t_2$,
perpendicular to => **dangerous line**
through P_1, P_2,
P is unstably determined

Example: Determination of point P by two intersecting circles defined by measure-
            ment of two distances s_1, s_2
⊛   given points P_1, P_2
○   new point P
t_1, t_2 tangents in P at circle 1 or 2, respectively

Fig. 2.10. Critical configuration of the second kind: instability.

point P without changing the observations s_1, s_2. These features characterize
a critical configuration of the second kind. In the presence of noise a solution
will generally be computable, but it reveals itself to be too sensitive with
respect to noise (random observation errors of s_1, s_2 and rounding errors): A
small change of second order magnitude of the observable distances s_1, s_2 will
result in a large, first order magnitude change of the position of the new point
P. Again, this type of critical configuration is not acceptable for applications.

We mentioned earlier that some orientation approaches are transformed
into polynomials for which a solution can be given in closed form. Now, with
the polynomial approach similar behavior of instability can be observed. Dis-

cretely disjoint solutions on the abscissa of the polynomial will converge to one multiple solution if a critical configuration is approached. In the limit there is one solution on the abscissa with its tangent identical with the abscissa. The geometrical similarity with the example of Fig. 2.10b is obvious.

Critical configurations of the first and second kinds have to be dealt with the relative orientation of two images and with the space resection of one image (Sects. 2.4.3, 2.4.4).

Finally, in our preparatory discussions on critical configurations we have to be aware that orientation data are derived from corresponding points which are always stochastic variables, i.e. they have to be associated with a region of uncertainty, e.g. defined by standard deviations or by an equivalent error ellipse. Consequently, to the results of orientation computation we have to attach a similar region of uncertainty determined by error propagation, see the comparable examples of Figs. 2.10a,b. Then it is easy to decide whether a new point P is at a statistically safe distance from a critical configuration (Fig. 2.10a) or not (Fig. 2.10b). Now, it is also imaginable to construct with a similar error propagation as before an accompanying region in the neighborhood of a critical line or surface (denoted as a dangerous region or a dangerous space section). Configurations of points inside these sections will have the same critical impact on the orientation solution as if all relevant points are exactly positioned on that danger line or surface. Considerations of that kind have already been conveyed in the past and solutions have been presented: in photogrammetry by Killian [8] and in projective geometry by Krames [109].

2.4.3   Uniqueness Conditions of 2D-2D Relative Orientation

Among the orientation tasks relative orientation plays a particular part in so far as orientation has to be obtained only from relationships between two bundles of imaging rays without any metric information about the imaged object. Therefore, it is not surprising that here the number of solutions may be larger and the conditions of uniqueness are more complex than with the other orientation tasks. Nevertheless, for relative orientation there exists a complete mathematical theory solving all its uniqueness problems with synthetic projective geometry in extreme generality. First impulses to investigate uniqueness of relative orientation evolved from applications of aerial photogrammetry. Characteristic shapes of valleys in the Alps gave rise to evident problems of relative orientation. First results were achieved with analytical geometry by photogrammetrists: the Swiss R. Bosshardt 1933 ("dangerous cylinder") and the Germans R. Finsterwalder, 1934, 1938 and E. Gotthardt [7] (main types of "dangerous quadrics"). The synthetic investigations of uniqueness of relative orientation were started by the French photogrammetrist E.C.P. Poivilliers 1937. Independently of him the Austrian mathematicians Krames 1937 and Wunderlich 1941 [6] have clarified all questions of uniqueness with full mathematical rigor and generality. Many supplementary investigations followed giving details for practical work with photogrammetric stereo compilation machines and for statistical error analysis, see the nice dissertation of Hofmann [9] and the very good survey of Buchanan [111] over that part of projective geometry in photogrammetry since its beginnings. The computer vision community rediscovered that subject in the 1980s, at first without discussing the results of Krames and Wunderlich, later profound and modern treatments (preferring algebraic projective geometry) were presented, e.g. by Maybank 1993 [63], Faugeras 1993 [28] and many others.

It is one of the proper aims of projective geometry to analyse and synthesize curves and surfaces in 3D space as intersection products from simple basic geometrical elements, like lines, planes, stars of rays, pencils of planes and so on (see e.g. [73]). Krames [5,112,113] follows that idea. He interprets the stars of rays of two images with given interior orientation (Table 2.1, (2.2)) but still unknown relative orientation as two pencils of congruent planes. The axis of the pencil of planes from the first bundle of rays is the (fictitious) imaging ray to the second center of image rays (= epipolar axis). Each image ray of the first bundle together with the epipolar axis defines one plane (= epipolar plane) of the first pencil of planes. The second pencil of planes is defined accordingly. The two pencils of planes are congruent, because planes of corresponding imaging rays have to coincide when relative orientation is established. Therefore, the angles between corresponding planes of the two pencils are equal. Krames investigated in detail how many times and under which conditions the corresponding imaging rays of both pencils of planes will intersect, and thus will be in relative orientation, and what kind of space surfaces will thereby be obtained from the intersections of the corresponding planes. In Sect. 2.3.1 similar findings for the relative orientation of projective collinear stars of rays were reported. However, now we have to orient perspective collinear stars of rays. Again surfaces of second degree show up to be the only critical configurations; however, much more specialized than before as explained below:

(1) Finsterwalder [2,3] proves that the relative orientation of two images can be established by intersection of two stars of rays from five pairs of corresponding image points. Kruppa [4] investigates the relative orientation with respect to the number of solutions. He states that with five pairs of image points there are at most eleven solutions (some of them are imaginary) if the same number of so-called supplementary orientations is excluded. Faugeras and Maybank 1989 [110], p. 19, have shown in their important paper that two of the eleven solutions always coincide, so that Kruppa's upper bound reduces to only *ten solutions of relative orientation*.

(2) *Supplementary solutions of relative orientation* can easily be detected and excluded if only the intersections of corresponding rays in front of the camera are admitted which is evident in practical application ([5], p. 331).

Consequently, image rays have to be used as half-rays only, defined by
a fixed sequence within which a ray passes through the image point,
the center of perspective and object point (Fig. 2.11). Of course, this
definition may be useful as a check of an orientation.

Besides, the two stars of image rays of a supplementary orientation will
intersect and yield a nonattractive "deformed object surface", being in a
harmonically collinear relationship to the surface in front of the images.
Usually a similarity relationship of a reconstructed object surface with
respect to the true object surface is desired. In the following we do not
count any more supplementary solutions.

(3) Krames [5], p. 346, proves that with five or more pairs of corresponding image points from two images *two or three real solutions of relative
orientation exist if and only if the corresponding object points together
with the centers of perspective of the two images belong to one of certain orthogonal ruled quadrics*, specified in (5). The solutions generally
are distinct, but may coincide. Solutions leading to object surface reconstructions which are in mutual similarity relationship are excluded here.
There may be more than three (or even infinitely many) solutions if the
quadratic relationship between both images is not unique (see (6) and
(7)).

(4) Krames [5], p. 347, further proves that *with six or more pairs of corresponding image points from object space points not belonging to one of
certain orthogonal ruled quadrics*, specified in (5), the *relative orientation
is uniquely determined*.

(5) Two real congruent pencils of planes generate specific ruled quadrics, the
so-called *orthogonal ruled quadrics* ([5], p. 337). These are all orthogonal
cones, all orthogonal hyperboloids of one sheet (Fig. 2.12), the orthogonal
hyperbolic paraboloids, the circular cylinders and the equilateral hyperbolic cylinders. Additionally, there are the pairs of orthogonal planes as
degenerate ruled quadrics. Object points on these surfaces will give rise
to multiple solutions of relative orientation not before these surfaces are
in a specific position relative to the real centers of projection O_1, O_2,
(Fig. 2.12). The axes e_1, e_2 of two congruent pencils of planes, which
generate the surface, have to be so-called adjoint generating lines (they
are interchanged when rotated by 180° about the longest diameter of the
hyperboloid) and have to pass through O_1 and O_2. There are ∞^1 pairs
of adjoint generating lines, two of them, e_1, e_2, are depicted in Fig. 2.12.
Besides, there are also in general four coinciding pairs of adjoint generating lines, which are called principal generating lines, see e_s in Fig. 2.12.
It still acts as the axis of pencils of planes for the generation of the
quadric and on which the centers of projection O_1, O_2 may be located.
This case is shown in Fig. 2.12. If O_1, O_2 are located on distinct adjoint

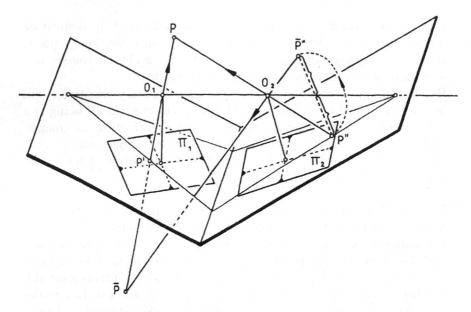

π_1, π_2	two images in relative orientation
P	object point in front of the two images (cameras), reconstructed from corresponding image points P' and P''
$\overline{P}''$	position of image point P'' after rotation by 180° of image π_2 about the epipolar axis through the centers of perspective O_1, O_2
$\overline{P}$	object point reconstructed from $P', \overline{P}''$ in supplementary relative orientation

Fig. 2.11. Supplementary solution of relative orientation.

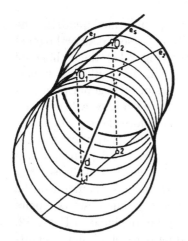

e_1, e_2 adjoint generating lines and axes of two congruent pencils of planes for generation of the hyperboloid (the planes are not shown)

e_s principle generating line with centers of perspective O_1, O_2 of two images

d diameter of hyperboloid

Fig. 2.12. Critical configuration of 2D-2D relative orientation: Orthogonal hyperboloid of second degree and one sheet.

generating lines, then there are up to two or three real and distinct solutions of relative orientation. On the other side these solutions coincide to one unstable solution (see Sect. 2.4.2), if O_1, O_2 are located on a principal generating line. The orthogonality property of these quadrics can be shown algebraically and geometrically. The equation of a hyperboloid related to its principal axes reads $(x^2/a^2) + (y^2/b^2) - (z^2/c^2) = 1$. Then the following relationship $(1/a^2) - (1/b^2) + (1/c^2) = 0$ is valid, where $a > b$ and c is the imaginary half-diameter of the hyperboloid. The geometrical consequence is such that cross-sections of parallel planes orthogonal to a principal generating line of these quadrics will produce circles (Fig. 2.12). An arbitrary quadric surface is determined by nine points or conditions; this is well known. Because of the abovementioned four conditions (orthogonality, $O_1 =$ surface point, $O_2 =$ surface point, $e_1 =$ adjoint e_2) that have to be fulfilled, there remains a five-dimensional manifold for the specific orthogonal ruled quadrics with which a unique solution of relative orientation cannot be achieved. These specific quadrics comprise: ∞^5 orthogonal hyperboloids, ∞^4 orthogonal cones or paraboloids, ∞^3 equilateral hyperbolae and circular cylinders and finally as degenerations all ∞^3 planes in 3D space (see also (7)). The relatively high dimension of these quadrics demonstrate that these critical surfaces are not so very special or rigid, so that natural surfaces and more than ever artificial surfaces (manmade surfaces are said to be composed of about 80% of quadrics) may easily belong to that group. When aerial images from the earth have to be evaluated the object plane is most likely to be expected and in mountain areas the "dangerous cylinder" or similar quadrics. Of the quadrics involved with relative orientation only hyperboloids and paraboloids can lead to up to three real solutions and only in rare cases. With all the other quadrics up to two real solutions are to be expected.

(6) The practical importance of multiple solutions of relative orientation depends on their geometrical relationships which exist between one another and in relation to the 3D metric object surface. It is clear a priori each orientation of the two bundles of imaging rays is well accepted as long as the intersections of corresponding rays form a surface and the reconstructed surface is geometrically similar to the true object surface.
Krames [5], p. 344, proves that of the solutions only one orientation (= primary solution) produces a surface that has the desired similarity relationship with the metric object surface. For the rest of the resulting surfaces (= secondary solutions of orientation) there exists a bijective (besides some rare exceptions) quadratic relationship with one another and with the true metric object surface. If the object surface is a general plane this relationship reduces to a collineation. Finally, Krames [112], p. 65, shows how to move the two stars of rays from one orientation to the other in both directions.

The geometrical relationships of the secondary solutions of relative orientation in photogrammetry so far have been regarded as really detrimental. Therefore, all quadrics that are the origin for multiple solutions are called by Krames [5] *"dangerous surfaces in the general sense"*. If relative orientation is performed interactively at a photogrammetric workstation a distinct secondary solution can be easily detected *just by stereo viewing of the operator* and can easily be eliminated. However, the instance when primary and secondary solution are close by or coincide, is not easy to detect just by viewing and thus the risk involved is greater than before. Consequently in the past the name *"dangerous surfaces in the closer sense"* has been given to these object surfaces. In Sect. 2.4.2 as we have already clarified such orientation results are very unstable or in rare cases may even be infinitely ambiguous. Anyway, these orientations are not acceptable for application.

(7) *Plane object surfaces* presumably represent the most common surfaces when applying stereo vision. So, some more information is given on relative orientation of images with respect to that type of dangerous surface. The outline of a direct computational algorithm for its two solutions has already been reported in Sect. 2.3.1. Now, the following configurations of stereo images and the resulting relative orientations are interesting ([5], p. 332 ff):

(a) Both camera positions O_1, O_2 are on the same side of a general object plane ε. We can always add the specific epipolar plane $\bar{\varepsilon}$ (passing through O_1, O_2) being orthogonal to plane ε. This pair of planes $\varepsilon, \bar{\varepsilon}$ represents a degenerate ruled quadric. It is extremely unlikely that there are really object points on $\bar{\varepsilon}$. Besides the primary solution of relative orientation we will have the secondary solution where its centers of perspective O'_1, O'_2 are positioned on two distinct generating lines e_1, e_2 in the epipolar plane $\bar{\varepsilon}$. e_1, e_2 are the axes of two congruent pencils of planes which generate this orthogonal pair of planes $\bar{\varepsilon}, \varepsilon_1$. e_1 is defined by O_1 and the mirror image of O_2 with respect to ε, and e_2 lies symmetrically to e_1, also with respect to ε. In this case (a) the solutions of relative orientation are very different and, therefore, there is the possibility of detecting the correct one with image analysis tools. Finally, the solution will be stable as long as the four image points, necessary for orientation, are positioned all over the field of view of the cameras.

(b) Both solutions of relative orientation coincide into one yet unstable orientation if both centers of projection, O_1 and O_2, are positioned on (or are close by) a normal to the object plane ε. This relative orientation is not acceptable.

(c) In few cases stereo images of plane objects may produce two congruent stars of imaging rays. In these rare cases the number of solutions of relative orientation is indefinite ([5], p. 346). Example 1: The centers of perspective are positioned symmetrically on different sides of the object plane. There are ∞^2 solutions of relative orientation. Example 2: The object surface is the plane ε at infinity, corresponding imaging rays are parallel. Therefore, on their connecting line the centers of perspective can translate as many as ∞^1 times. Finally, we have to mention two special yet trivial cases of the two orthogonal planes as critical configurations: these cases are present when all object points are positioned on one line or on two skew lines.

By the way, it is interesting to notice that the critical configurations from this subsection (c) for relative orientation with two congruent stars of rays at the same time are critical for the 2D-3D orientation of a single image, which is quite obvious, see Sect. 2.4.4.

After collecting the complete uniqueness conditions of relative orientation we now have to discuss how to deal with them in practical applications. Up to now our presentation of computational procedures of relative orientation (Sect. 2.3.1) does not comprise elements of self-diagnosis whether a computed solution is the correct one among the possible two or three real solutions. Yet, following [112], p. 65, it is possible to proceed from one computed solution to the next.

In classical photogrammetry, greatly assisted by an operator, there are simple practical procedures to overcome the problems of multiple solutions, in use for more than fifty years [9]. They are based on the theoretical findings given above. The technical and conceptual means of present automated digital photogrammetry and of computer vision have changed this situation completely. But, in principle there are the same approaches for success, namely, either by an a posteriori analysis of the orientation parameters with subsequent remedy by new information or by a priori precautions to prevent the outcome of multiple solutions from the beginning.

Let us now disregard the extremely rare cases of complete indeterminacy. Then the parameters of relative orientation can always be estimated together with its variance–covariance matrix (Gauss–Markov estimation) even though in some cases the appertaining equations may be very ill-conditioned. In this respect the redundancy of the system of equations is allowed to be zero (minimum solution) or greater than zero.

After orientation it is easy to compute object space points from corresponding image points by intersection in 3D space in an image centered coordinate system, see Sect. 2.3.1. Now, by computation it can be clarified whether the 3D space points and the two centers of perspective are positioned too close to one of the ruled quadrics specified in (5). The orientation is uniquely determined if at least one point at high statistical significance is distinct from all these danger surfaces. If all points are in conformity with one of the dangerous quadrics then it is clarified that there are two solutions or even three depending on the type of best approximated quadric. In the next step a new attempt can be started to solve the relative orientation for the correct and stable solution simply by searching for additional corresponding feature points in the images and by subsequently repeating the orientation computation. Anyway, by additional image information the probability of getting trapped in a critical configuration of a scene will be substantially reduced. This strategy will always succeed, if a noncritical object surface exists at all. However, if in fact all object points and all perspective centers belong to a danger quadric then with arbitrarily many corresponding image points no unique or stable solution can be achieved (see (3) and (6) in this section). In that case independent metric object surface information can be integrated into the examination of the solutions, for example space distances, horizontal or vertical lines or planes, 3D reference points, and 3D space coordinates of centers of perspective (e.g. from the Global Positioning System, Fig. 2.4).

Suppose there is a minimum of redundancy. Then it can be definitely and readily clarified whether the reconstructed object surface is in a similarity relationship with the metric object space. If this is true, then the correct primary solution has already been computed, otherwise, according to [112], p. 65, we may step over to the next solution and continue the examination.

Although the above outlined procedure is feasible the following approach would be much more elegant: circumvention of relative and absolute orientation by a unified procedure in the sense of closed-form bundle block formation (Table 2.2, no. 5). The iterative least-squares version of it (Gauß–Markov estimation) exists for a long time; however, there is the burden of procuring automatically accurate enough starting values for the nonlinear parameters. Of course, a general direct closed-form solution algorithm would be preferred without doubt. Apparently, it does not yet exist. With that procedure all the aforementioned rather lengthy problems of multiple or unstable solutions of relative orientation are skipped from the beginning with just one trivial exception, namely that all object points are located on one line.

2.4.4 Uniqueness Conditions of 2D-3D Image Orientation (Space Resection)

As for relative orientation there exists also for 2D-3D image orientation a complete and coherent mathematical theory about its uniqueness conditions. These issues have been derived for the projective and for the perspective collineation, as well. They are so closely related that both will be discussed in this section. In photogrammetric and computer vision references often isolated examples of critical configurations of reference points (including the center of perspective) have been investigated by Finsterwalder [2,3], Gotthardt [114,116], Killian [87], Thompson 1966 [117], and Fischler and Bolles [38]). In this section we want to sum up the issues under a unified and complete theory. All fundamental investigations, proofs and results were obtained with synthetic projective geometry in the 19th century, see especially Sturm [73]. Everything has been found while studying the intersection of two collinear and two congruent stars of rays, originating from the projective and perspective camera model, respectively (Table 2.1). A very instructive historical review and summary about the uniqueness conditions for orientation with projective collineation is given by Buchanan [118]. Equivalent information with respect to the perspective collineation has been presented by Wunderlich [6] and Killian [120].

2.4.4.1 Uniqueness Conditions of Image Orientation with Projective Collineation.

Before we start with uniqueness conditions depending on the configuration of reference points we first have to ensure that the computational procedure of the parameters of the projective camera model (Table 2.1, (2.1)) used under general and favorable configurations is free from singularities. Similar reasoning has already been concerned with the perspective camera model, see Sect. 2.4.1 on singularity free representation of rotation. The determination of all 12 elements of the 3×4 matrix $A = \{a_{ij}\}$ from (2.1) with six corresponding points is not unique since these equations are homogenous. To fill up the rank defect usually a constraint, e.g. $a_{34} = 1$, is introduced, i.e. all elements a_{ij} of the matrix A are determined up to a specific scale factor. However, this choice is only sucessfull as long as a_{34} itself is nonzero, which depends on the position and orientation of the object coordinate system with respect to the image reference frame. A general and singularity free solution is obtained with the alternative constraint $a_{31}^2 + a_{32}^2 + a_{33}^2 = 1$ ([28]). It's computational cost can even be lower than with the constraint $a_{34} = 1$ ([14,15]). In the following we suppose the application of the alternative constraint so that no restrictions on the coordinate systems are imposed.

Now, all geometric loci in object space are given on which the six or more reference points and the center of projection are not allowed to lie on if a unique solution of orientation is desired. If the points are on a dangerous locus then the linear equation system (2.1), Table 2.1, will have a rank defect. Accordingly, the solution is represented by a subspace within the space of the eleven parameters of orientation, but not by a solution point. However, if the reference points are sufficiently distant from these critical surfaces or lines the solution will be unique.

(a) Given an image with six or more image points it is easy to construct with respect to them different yet collinear stars of rays. Consequently, several collineations must exist also as the relationship of image points with corresponding reference points in 3D space if the reference points all lie on a plane and on a line which passes through the center of projection. This is already the first critical configuration with which a unique orientation cannot be obtained.

(b) *Space curve of third order.* Again we consider two different yet collinear stars of rays originating from the points of one image. To each ray of one star belongs a corresponding ray from the other star. Corresponding rays are related to the same image point. In 1827 Seydewitz ([73], vol. II, p. 173) was the first to show that among the ∞^2 pairs of corresponding rays of two collinear stars there are only ∞^1 pairs that intersect in 3D space; all the other corresponding rays are skew. The geometric locus of the intersections is an arbitrary twisted space curve of third order or a degeneration of it, see (c) and (d). All these curves have to pass through the centers of projection, as well.

It is immediately evident that the twisted cubic is not acceptable as a locus for reference points because more than one bundle of projective rays can be oriented to it. A possibly composite twisted cubic passes through any six reference points in space. Thus six points (this is also the minimum number for image orientation with projective collineation) cannot ensure unique image orientation over all 3D space because when taking an image from a general position it is not known whether the camera will be located on that cubic or not. Of course, ambiguous solutions exist only if the camera lies also on the cubic!

(c) *Degenerate third order space curve I.* As first degeneration of the cubic we mention the conic together with a unisecant. The unisecant intersects the conic and passes through the center of projection. This configuration is present e.g. if four reference points lie on a plane forming a conic (ellipse, hyperbola, ...) and two are on the unisecant.

(d) *Degenerate third order space curve II.* A stronger degeneration consists of two skew lines and a transversal, intersecting both lines and passing through the center of projection. The points of intersection may lie on the plane at infinity.

We complete the given report by naming some trivial critical configurations. These are specialties of the cases already given, namely all reference points lie only on one plane or only on one conic or only on one line or are joined in only one point.

Let us summarize the uniqueness conditions of image orientation with projective collineation: With the minimum number of six reference points there is no guarantee of a unique image orientation over all 3D space, if the projective camera model is used. The presented critical configurations (a)–(d) may occur with practical application without doubt, especially in indoor scenes, urban scenes or with industrial objects. A simple increase of the number of reference points does not force a guarantee for uniqueness. On the other hand if more than six reference points are given "in a general position in object space", i.e. well distributed over the image frame, no three points are collinear and all reference points together with the center of projection are not close to one of the critical configurations (a)–(d), then image orientation is unique. Two examples are given: First, seven reference points are distributed over two distinct planes. Second, and similarly, the reference points show up differences in the viewing direction of the camera of about one third of the mean distance between the camera and the scene.

2.4.4.2 Uniqueness Conditions of Image Orientation with Perspective Collineation. In contrast to the projective collineation camera model, the uniqueness conditions with the perspective camera model are more specific and thus simpler. Now, a star of imaging rays has to be related to the set of reference points with five parameters less (six instead of eleven, see Table 2.1, (2.2)), thus a star of rays is now more rigid, because the data of the interior orientation of an image are given and are kept fixed. If there is more than one orientation of a star of image rays with respect to three or more given reference points, then these stars have to be congruent and not only projectively collinear as with the stars in Sect. 2.4.4.1. Therefore, there exist per se fewer space positions of the reference points than before that are critical.

Space resection when performed with the minimum number of three reference points will lead to up to four real solutions (Sect. 2.3). Now, we want to interpret this important finding geometrically. As a matter of fact there are in general altogether up to eight discrete solutions, four of which are positioned symmetrically on either side of the plane through the three given points P_1, P_2, P_3 (Fig. 2.8).

This can be seen by the following geometric solution for space resection, namely by intersection of three toroids (Monge 1798 [84], Grunert [85]). Each toroid is defined by one of the face angles ψ_{12}, ψ_{13}, ψ_{23} and related to s_{12}, s_{13}, s_{23}, respectively, as axes (Fig. 2.8). However, only the solutions from one side of the plane through P_1, P_2, P_3 can be realistic. The solutions of the wrong side can easily be excluded by using all image rays as half-rays only, passing through the relevant points in a fixed sequence, chosen once for all. In Fig. 2.8 the image is defined in the position of a photographic negative for which the half-rays are fixed by the following sequence: image point, then center of perspective, then object point. So, the three toroids will intersect in up to four discrete solution points. One obvious example: The base triangle with vertices P_1, P_2, P_3 is equilateral, $s_{12} = s_{13} = s_{23}$, and also the three legs of the tetrahedron are all equal, $s_1 = s_2 = s_3$. Then there are four real solutions of space resection which are positioned as follows: the first solution is on the vertical line erected in the center of the circumscribed circle of the base triangle P_1, P_2, P_3, the second solution is on leg 1 of the tetrahedron as defined by the first solution with P_1, P_2, P_3 and so on for the third and fourth solution symmetrically positioned on the same tetrahedron.

Now, it is important to know which are the complete conditions for multiple solutions. All critical configurations of reference points with which multiple distinct or coinciding (unstable) solutions are unavoidable are known [73]. The configurations will be given in analogy to those of Sect. 2.4.4.1. Again, there are two groups, group (a) related to a plane, and (b)–(e) related to a cubic curve, however, more specific and therefore rarer with real scenes than before in Sect. 2.4.4.1.

(a) Image orientation based on the projective collineation is always ambiguous if all reference points are on an object plane ((a) of Sect. 2.4.4.1). In comparison with the perspective collineation, image orientation with four or more reference points on a general object plane always results in a unique and stable solution (Gotthardt [114], Wunderlich 1943 [115]), if extreme cases like the following two are avoided. The first case can be regarded as a specialty of Sect. 2.4.4.1 (a): Image orientation with perspective collineation is unstable (coinciding solutions), if three or more reference points together with the center of perspective are positioned on one object plane [114,117]. There is another special case of the object plane which comes out as a specialty of the horopter curve, see (e) below. In this case the center of perspective and the object plane are separated by a long distance.

(b) *The horopter curve on the dangerous cylinder.* The ∞^2 corresponding rays of two congruent stars of rays in general position do not intersect with the exception of ∞^1 pairs. The points of intersection again are generating – in analogy to Sect. 2.4.4.1(b) – a third order twisted space curve, however, a very special one [73]. This cubic also passes through the vertices of the stars of rays but is located on a circular cylinder. This is the "dangerous cylinder of space resection", well-known in photogrammetry since the time of Finsterwalder [2]. The special cubic has also been studied by Helmholtz [73] in connection with investigations in physiological optics about the geometry of stereo vision (= locus of space points which are seen distinct) and named by him the "horopter" (in projective geometry also called the "right cubic circle"). There are detailed investigations about the horopter and its degenerations; apparently the first originate from Möbius, 1827 [73].

If there are in all – besides case (a) – multiple distinct or coinciding solutions of image orientation then all reference points (three or more!) and all solutions have to be positioned on the horopter or on one of its degenerations. This statement is confirmed by a theorem of Schur 1889 [119], according to which the same horopter can be generated not only by a single congruent pair of stars of rays but by ∞ pairs, the vertices of which all lie on the horopter symmetrically to the horopter's vertex. With reference to the special coordinate system of Fig. 2.13 the analytical equations of the horopter read [120]:

$$x = k\sqrt{z}/\sqrt{a-z}\,; \qquad y = \sqrt{z(a-z)}\,; \qquad 0 \le z \le a\,, \qquad (2.20)$$

where k is the parameter of the horopter and a the diameter of the circular cylinder. The horopter arrives from ∞ on one side of its asymptote (= x-axis), winds around the cylinder in a right-handed screw motion, then approaches the asymptote from its other side and finally vanishes at infinity. Equations (2.20) are well suited for an examination of the given reference points as to whether they are eventually on a circular cylinder or even on a horopter. It can be shown that a horopter with its cylinder in general position in 3D space is uniquely determined by four given reference points, and that ∞^2 horopter curves pass through any three reference points in 3D space, which is also the minimum number of points for the solution of space resection. It has to be stressed that its solution is ambiguous if and only if the center of perspective of the camera is located also on one of those horopters. All one to four solutions, whether distinct or not, are on the same horopter.

In the past in most references only the dangerous cylinder has been discussed, not recognizing the proper role of the horopter curve. Very revealing and not really surprising are the close relationships of the horopter with the ambiguous solutions of relative orientation (Sect. 2.4.3). With every solution of relative orientation we have to associate an individual orthogonal ruled quadric. Now, their intersection curve defines a horopter, see Wunderlich [6] and Krames [5].

As before in Sect. 2.4.4.1 there are degenerations of the horopter resulting in the simplest critical configurations of space resection [73,6].

(c) *Degenerate horopter I.* When the horopter parameter k converges to zero, see Fig. 2.13, the horopter degenerates into a circle and a straight line. This particular horopter is related to two coinciding solutions of space resection. Therefore we have an unstable result ([116]).

(d) *Degenerate horopter II.* There is a specialty case of Sect. 2.4.4.1(d): The center of perspective lies on the transversal which is normally intersected by two lines. Evidently this is a rare case.

(e) *Degenerate horopter III.* With human stereo vision very natural, with images from cameras however very seldom, we have the following special case, compare (a): If the reference points are all on the plane at infinity,

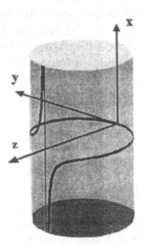 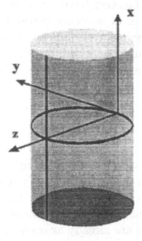

Fig. 2.13. Critical configuration of 2D-3D image orientation: the horopter curve on the circular cylinder; on the left parameter $k = 0.5$, on the right $k = 0$.

then the horopter degenerates into the plane at infinity and the line joining the centers of congruent stars of rays. Corresponding rays are now parallel. Both cases where the reference points are on a plane, (a) and (e), in practical application can easily be avoided.

As before in Sect. 2.4.4.1 again here are some trivial critical configurations of reference points which are at the same time special cases of (a)–(e): the circle (see Fig. 2.9 and Fig. 2.13), the line, and the point.

Now, also for Sect. 2.4.4.2 above, the uniqueness conditions of image orientation with perspective collineation can be formulated. We have seen that altogether the conditions are similar to those of projective collineation, yet by far more favorable in practical applications:

- With the minimum number of three reference points for space resection there always exists one of the configurations (a)–(e) which becomes critical disregarding the position of the camera. Therefore, in this situation there is no guarantee that image orientation is uniquely feasible all over 3D space. Most of the critical configurations will be met very rarely in practical applications or can easily be detected a priori. Only the horopter (= locus of all reference points and of all solutions) may have some importance.
- With the choice of four reference points in particular positions the formation of a horopter a priori can be prevented. Example: With four reference points in an object plane, where no three of them are collinear, and with the plane not passing through the center of perspective, there always exists a unique solution.
- The increase in the number of reference points alone (more than three) does not give any guarantee of obtaining a unique and stable solution. On the other hand if the reference points are in general position in 3D space, i.e. at a secure distance from the configurations (a)–(e) and no three points are collinear, then image orientation will be without problems: In any case it will be unique and – if the reference points are spread out all over the image format and also in the corresponding object space region – orientation will be stable, as well.

2.5   Conclusion

A review has been given of the minimum solutions of various orientation tasks of photogrammetry and computer vision. This review has been based on examining the latest developments and also by looking back at the early beginnings of orientation procedures of photogrammetry to demonstrate that there are already answers to conceptual questions of today. Among the orientation tasks using corresponding points, three basic ones, 2D-2D relative orientation, 2D-3D image orientation and 3D-3D absolute orientation have

been discussed in detail. For all of them closed-form direct solution procedures are available although at different levels of compactness.

Due to the importance for automation of orientation, particular attention
has been focussed on the representation of rotation, free from singularities in
full parameter space, and on the critical configurations of points in 3D object
space, i.e. center(s) of perspective and points on an object surface. Existence
and uniqueness of orientation solutions directly depend on both. Complete
uniqueness conditions of the three basic orientation tasks were derived long
ago, as part of basic research work in projective geometry, namely in the 19th
century and in 1930–1948. The importance of projective geometry for vision
geometry, not only from its beginnings for photogrammetry but today also
for computer vision cannot be overestimated.

Apparently, a renaissance in projective geometry is now taking place.
Indeed, improved procedures of orientation are desirable if you summarize
the rather restrictive uniqueness conditions of 2D-2D relative and 2D-3D
single image orientation. Closed-form solutions of combined relative and absolute orientation (= bundle block formation, see Table 2.2) and a similar
combined relative orientation of more than two images (all with perspective
collineation) will completely remove or drastically reduce the risk of critical
configurations. The same is true for the specific properties of multi-image
configurations based on projective collineation. Again, these multi-image orientation tasks can be attacked with sucess by projective geometry. And this is
what actually takes place at the present time, see again Hartley [76], Shashua
and Nabab [30], Zeller and Faugeras [67], also Niini [68,32], Heyden [61] and
Hartley [60].

At the end of this review two general questions about the future trends of
image orientation procedures in photogrammetry and computer vision should
be allowed: Will image orientation still be solved with predetermined corresponding features, like points and lines? And will camera orientation continue
to be based on image processing? In both directions, I believe, a change may
occur within the next decade. Regarding the second question, it should be
noticed that the orientation of aerial images is already substantially eased
and automated by integration of 3D positional and attitude sensors into orientation (like GPS, Fig. 2.4, and INS). Primarily, it is a question of accuracy
or cost that the same will soon happen in close-range photogrammetry; however, there always will be types of application (e.g. in closed rooms) where
orientation has to be performed by image processing.

Finally, as for the first question, we now rely on sufficiently strong, representative yet specially extracted image features. However, there are proposals
(see Sect. 2.2) to integrate all grey values and its derivatives into image orientation although at the moment the computational expense is high. So, there
still exists a strong potential for ongoing improvements of image orientation
approaches and, therefore, it will not cease to be an interesting mathematical
and computational problem in the next decade, as it is today.

References

1. J.L. Lagrange. Lecons éléementaires sur les mathématiques, données à l'École Normale en 1795. In M.J.-A. Serret (ed.): *Oeuvres de Lagrange.* Tome 7, Section IV, Gauthier-Villars, pp. 183–288, Paris, 1877.

2. S. Finsterwalder. Die geometrischen Grundlagen der Photogrammetrie. Jahresbericht Deutsche Mathem. Vereinigung, VI, 2, Teubner Verlag, Leipzig, pp. 1–41, 1899. In *Sebastian Finsterwalder zum 75. Geburtstage,* Deutsche Gesellschaft für Photogrammetrie, Wichmann Verlag, Berlin, 1937.

3. S. Finsterwalder. Eine Grundaufgabe der Photogrammetrie und ihre Anwendung auf Ballonaufnahmen. Abhandlung Königlich-Bayerische Akademie d. Wissenschaften, II. Kl., XXII. Bd., II. Abt., pp. 225–260, 1903. In *Sebastian Finsterwalder zum 75. Geburtstage,* Deutsche Gesellschaft für Photogrammetrie, Wichmann Verlag, Berlin, 1937.

4. E. Kruppa. Zur Ermittlung eines Objekts aus zwei Perspektiven mit innerer Orientierung. Sitzungsberichte Akad. Wiss. Wien, Math. Naturw. Kl. (Abt. IIa. 122), 1939–1948, 1913.

5. J. Krames. Zur Ermittlung eines Objektes aus zwei Perspektiven. Ein Beitrag zur Theorie der "gefährlichen Örter". Monatshefte für Mathematik und Physik **49**, 327–354, 1941.

6. W. Wunderlich. Zur Eindeutigkeitsfrage der Hauptaufgabe der Photogrammetrie. Monatshefte für Mathematik und Physik **50**, 151–164, 1941.

7. E. Gotthardt. Der gefährliche Ort bei der photogrammetrischen Hauptaufgabe. Zeitschrift für Vermessungswesen **10**, 297–304, 1939.

8. K. Killian. Über die bei der gegenseitigen Orientierung von Luftbildern vorkommenden gefährlichen Flächen und "gefährlichen Räume". Photographische Korrespondenz **81**, no. 1–12, 13–23, 1945.

9. W. Hofmann. Das Problem der "Gefährlichen Flächen" in Theorie und Praxis. Ein Beitrag zur Hauptaufgabe der Photogrammetrie. Doct. Thesis, Deutsche Geodät. Komm., Reihe C, **3**, München, 1953.

10. F. Hohenberg, J.P. Tschupik. Die geometrischen Grundlagen der Photogrammetrie. In K. Rinner, R. Burkhardt (eds.): *Handbuch der Vermessungskunde,* Bd. III, a/3 Photogrammetrie, pp. 2236–2295, Metzlersche Verlagsbuchhandlung, Stuttgart, 1972.

11. Y.I. Abdel-Aziz, H.M. Karara. Direct linear transformation from comparator coordinates into object space coordinates in close-range photogrammetry. In *Proc. ASP/UI Symposium on Close-Range Photogrammetry,* pp. 1–18, Urbana, Illinois, January 1971.

12. D.H. Ballard, Ch.H. Brown. Computer Vision. Prentice-Hall, Englewood Cliffs, New Jersey, 1982.

13. T.Y. Shih, W. Faig. Physical interpretation of the extended DLT-model. Proc. ASPRS Fall Convention, American Society for Photogrammetry and Remote Sensing, Reno, Nevada, pp. 385–394, 1987.

14. T. Melen. Geometrical modelling and calibration of video cameras for underwater navigation. Dr.Ing. thesis, Trondheim, Institut for Teknisk Kybernetikk Techn. Report 94–103-W, 1994.

15. T. Melen. Decompostion of the Direct Linear Transformation (DLT) matrix with symmetric representation of affine distortion. In: B. Olstad et al. (eds): Proc. of the NOBIM conference, pp. 111–122, Asker, Norway, June 1994.

16. W. Förstner. Determining the interior and exterior orientation of a single image. Report, Institute for Photogrammetry, University Bonn, 1999.
17. H. Fuchs. Projektive Geometrie. Anwendungen in Photogrammetrie und Robotik. Mitteilungen Geodät. Inst., Techn. Univ. Graz, Folge **63**, 1988.
18. H. Haggrèn, I. Niini. Relative orientation using 2D projective transformations. The Photogrammetric Journal of Finland **12(1)**, 22–23, 1990.
19. G. Brandstätter. Zur relativen Orientierung projektiver Bündel. Zeitschrift f. Photogrammetrie und Fernerkundung **6**, 199–212, 1991.
20. G. Brandstätter. Notes on the direct projective transformation of general stereo pairs into the rigorous normal case by image correlation. Intern. Archives of Photogrammetry and Remote Sensing **29**, part B3, Commission III, ISPRS, Washington, 1992.
21. G. Brandstätter. Fundamentals of algebro-projective photogrammetry. Sitzungsberichte Österreich. Akademie d. Wissensch., mathem.-naturwiss. Klasse, Abt. II **205**, 57–109, 1996.
22. K. Rinner. Studien über eine allgemeine, voraussetzungslose Lösung des Folgebildanschlusses. Österreichische Zeitschrift f. Vermessungswesen, Sonderheft **23**, 1963.
23. E.H. Thompson. The projective theory of relative orientation. Photogrammetria **23**, 67–75, 1968.
24. H. von Sanden. Die Bestimmung der Kernpunkte in der Photogrammetrie. Doct. Thesis, Univ. Göttingen, 1908.
25. H.C. Longuet-Higgins. A computer algorithm for reconstructing a scene from two projections. Nature **293**, 133–135, 1981.
26. J. Weng, P. Cohen, M. Herniou. Camera calibration with distortion models and accuracy estimation. IEEE Transactions on Pattern Analysis and Machine Intelligence **14(10)**, 965–980, 1992.
27. O.D. Faugeras. What can be seen in three dimensions with an uncalibrated stereo rig? In *Proc. 2nd European Conf. on Computer Vision*, Springer Verlag, Berlin, Heidelberg, pp. 563–578, 1992.
28. O.D. Faugeras. Three-dimensional Computer Vision. A geometric viewpoint. The MIT Press, Cambridge (Mass.), London (Engl.), 1993.
29. S.J. Maybank, O.D. Faugeras. A theory of self-calibration of a moving camera. Intern. Journal of Computer Vision **8(2)**, 123–151, 1992.
30. A. Shashua, N. Nabab. On the concept of relative affine structure. Zeitschrift für Photogrammetrie und Fernerkundung **5**, 181–189, 1994.
31. I. Niini. Relative orientation of multiple images using projective singular correlation. Intern. Archives of Photogrammetry and Remote Sensing **30**, part 3/2, 615–621, Comm. III, ISPRS, München 1994.
32. I. Niini. Photogrammetric block adjustment based on singular correlation. Acta Polytechnica Scandinavica, Civil Engineering and Building Construction Series no. 120, Espoo, 2000.
33. L. Quan. Invariants of six points and projective reconstruction from three uncalibrated images. IEEE Transactions on Pattern Analysis and Machine Intelligence **17(1)**, 34–46, 1995.
34. R.M. Haralick, L.G. Shapiro. *Computer and Robot Vision, Vol. I, II.* Addison Wesley Publ. Comp., Reading, Mass., 1992/1993.
35. G. Strunz. Bildorientierung und Objektrekonstruktion mit Punkten, Linien und Flächen. Doct. Thesis, Munich Univ. of Technology, 1992.

36. S.E. Masry. Digital mapping using entities: a new concept. Photogrammetric Engineering and Remote Sensing **48(11)**, 1561–1599, 1981.
37. J. Lugnani. The digital feature – a new source of control. Intern. Archives of Photogrammetry and Remote Sensing **24-III/2**, 188–202, Comm. III, 1982.
38. M.A. Fischler, R.C. Bolles. Random sample consensus: A paradigm for model fitting with applications to image analysis and automated cartography. Communications ACM **24(6)**, 381–395, 1981.
39. T. Schenk, Ch.K. Toth. Conceptual issues of softcopy photogrammetric workstations. Photogrammetric Engineering and Remote Sensing **58(1)**, 101–109, 1992.
40. H. Ebner, F. Müller. Processing of digital three-line imagery using a generalized model for combined point determination. Photogrammetria **41(3)**, 173–182, 1987.
41. D. Rosenholm, K. Torlegård. Three-dimensional absolute orientation of stereo models using digital elevation models. Photogrammetric Engineering and Remote Sensing **54(10)**, 1385–1389, 1988.
42. W. Schickler. Feature matching for outer orientation of single images using 3D wireframe control points. Intern. Archives of Photogrammetry and Remote Sensing **29**, Commission III, ISPRS, Washington, 1992.
43. V. Tsingas. Automatisierung der Punktübertragung in der Aerotriangulation durch mehrfache digitale Bildzuordnung. Doct. Thesis, Deutsche Geodät. Komm., Reihe C, no. 392, München, 1992.
44. B. Wrobel. Digital image matching by facets using objects space models. *Proc. Advances in Image Processing* SPIE – The International Society for Optical Engineering **804**, 325–333, The Hague, 1987.
45. B. Wrobel. The evolution of digital photogrammetry from analytical photogrammetry. Photogrammetric Record **13(77)**, 765–776, 1991.
46. M. Kempa. Hochaufgelöste Oberflächenbestimmung von Natursteinen und Orientierung von Bildern mit dem Facetten-Stereosehen. Doct. Thesis, Darmstadt Univ. of Technology, Institute for Photogrammetry and Cartography, 1995.
47. N. Ayache. Artificial vision for mobile robots. Stereo vision and multisensory perception. MIT Press, Cambridge, Mass., London, Engl., 1991.
48. L. Cogan, Th. Luhmann, S. Walker. Digital photogrammetry at Leica, Aarau. In Ebner, Fritsch, Heipke (eds.): *Digital Photogrammetric Systems*. Wichmann Verlag, Karlsruhe, 1991.
49. H. Haggrèn. On system development of photogrammetric stations for on-line manufacturing control. Doct. Thesis, Acta Polytechnica Scandinavica, Civil Engin. and Build. Constr. Series, Helsinki, **97**, 1992.
50. H.-G. Maas. Digitale Photogrammetrie in der dreidimensionalen Strömungsmeßtechnik. Doct. Thesis, Mitteilungen Institut für Geodäsie und Photogrammetrie, ETH Zürich **50**, 1992.
51. H. Schmid. Eine allgemeine analytische Lösung für die Aufgabe der Photogrammetrie. Bildmessung und Luftbildwesen, 103–113 und 1–12, 1958 und 1959.
52. Ch.C. Slama (ed.). *Manual of Photogrammetry*, 4th edition. American Soc. of Photogrammetry, Falls Church, VA, 1980.
53. A. Gruen. Photogrammetrische Punktbestimmung mit der Bündelmethode. Mitteilungen Institut für Geodäsie und Photogrammetrie, ETH Zürich, **40**, 1986.

54. L. Tang, C. Heipke. Automatic relative orientation of aerial images. Photogrammetric Engineering and Remote Sensing **62(1)**, 47–55, 1996.
55. J. Braun, L. Tang, R. Debitsch. PHODIS AT – An automated system for aerotriangulation. Intern. Archives of Photogrammetry and Remote Sensing **31(B2)**, 32–37, ISPRS, Wien, 1996.
56. E.M. Mikhail. Use of two-directional triplets in a sub-block approach of analytical aerotriangulation. Photogrammetric Engineering **29(6)**, 1014–1024, 1963.
57. V. Kratky. Present status of on-line analytical triangulation. Intern. Archives of Photogrammetry and Remote Sensing **23(B3)**, 379–388, Commission III, ISPRS, Hamburg, 1980.
58. A. Gruen. Algorithmic aspects in on-line triangulation. Photogrammetric Engineering and Remote Sensing **51(4)**, 419–436, 1985.
59. A. Gruen, Th. Kersten. Sequential estimation in robot vision. Intern. Archives of Photogrammetry and Remote Sensing **29**, Commission V, ISPRS, Washington, 1992.
60. R.I. Hartley. Computation of the quadrifocal tensor. In: H. Burkhardt and B. Neumann (eds.). Computer Vision – ECCV'98. 5th European Conference on Computer Vision, Freiburg, Germany, proceedings vol. I, 20–35, Springer Verlag Heidelberg, 1998.
61. A. Heyden. A common framework for multiple view tensors. In: H. Burkhardt and B. Neumann (eds.). Computer Vision – ECCV'98. 5th European Conference on Computer Vision, Freiburg, Germany, proceedings vol. I, 3–19, Springer Verlag Heidelberg, 1998.
62. T.S. Huang, O. Faugeras. Some properties of the E matrix in two-view motion estimation. IEEE Transactions on Pattern Analysis and Machine Intelligence **15(1)**, 1310–1312, 1989.
63. S. Maybank. Theory of reconstruction from image motion. Springer Verlag, Springer Series in Information Sciences **28**, Berlin Heidelberg, 1993.
64. J. Philip. A non-iterative algorithm for determining all essential matrices corresponding to five point pairs. Photogrammetric Record **15(88)**, 589–599, 1996.
65. J. Philip. Critical point configurations of the 5-, 6-, 7-, and 8-point algorithms for relative orientation. Report TRITA-MAT-1998-MA-13, Department of Mathematics, Royal Institute of Technology, Stockholm, 1998.
66. L. Hinsken. Algorithmen zur Beschaffung von Näherungswerten für die Orientierung von beliebig im Raum angeordneten Strahlenbündeln. Doct. Thesis. Deutsche Geodät. Komm., Reihe C, no. 333, München, 1987.
67. C. Zeller, O.D. Faugeras. Camera-self-calibration from video sequences: the Kruppa equations revisited. Rapport de recherche **2793**, INRIA, Sophia Antipolis, France, 1996.
68. I. Niini. Orthogonal 3-D reconstruction using video images. Intern. Archives of Photogrammetry and Remote Sensing **XXXI**, part. B3, 581–584, Vienna, 1996.
69. T. Melen. Extracting physical camera parameters from the 3×3 direct linear transformation matrix. In A. Gruen, H. Kahmen (Eds.): *Optical 3-D measurement techniques II. Applications in inspection, quality control and robotics*, pp. 355–365, H. Wichmann-Verlag, Karlsruhe, 1993.
70. J. Weng, Th. Huang, N. Ahuya. Motion and structure from two perspective views: algorithms, error analysis and error estimation. IEEE Transactions on Pattern Analysis and Machine Intelligence **11(5)**, 451–476, 1989.

71. M.E. Spetsakis, Y. Aloimonos. Optimal visual motion estimation: A note. IEEE Transactions on Pattern Analysis and Machine Intelligence **14(9)**, 959–964, 1992.

72. I. Hådem. Estimating approximate values before bundle adjustment in close-range photogrammetry – a review. In A. Gruen, E. Baltsavias (Eds.), *Proc. SPIE 1395*, pp. 1016–1027, 1990.

73. R. Sturm. *Die Lehre von den geometrischen Verwandtschaften*. 4 Bände, Teubner Verlag, Leipzig, Berlin, 1908.

74. W. Wunderlich. Rechnerische Rekonstruktion eines ebenen Objekts aus zwei Photographien. Mitteilungen Geodät. Inst. TU Graz **40**, (Festschrift K. Rinner zum 70. Geburtstag), 365–377, 1982.

75. H. Kager, K. Kraus, K. Novak. Entzerrung ohne Paßpunkte. Bildmessung und Luftbildwesen **2**, 43–53, 1985.

76. R.I. Hartley. Estimation of relative camera positions for uncalibrated cameras. In *Proc. 2nd European Conf. on Computer Vision, Springer Verlag*, Berlin, Heidelberg, pp. 579–587, 1992.

77. P. Stefanovic. Relative orientation – a new approach. Publ. Intern. Institute for Aerial Survey and Earth Sciences (ITC) **3**, 418–448, Enschede, The Netherlands, 1973.

78. C.M.A. van den Hout, P. Stefanovic. Efficient analytical relative orientation. Intern. Archives of Photogrammetry and Remote Sensing **21**, Comm. III, ISPRS, Helsinki, 1976.

79. G.H. Golub, C.F. van Loan. Matrix computations. North Oxford Publ. Co. Ltd., Oxford, 1983.

80. B.K.P. Horn. Relative orientation. Intern. Journal of Computer Vision **4**, 59–78, 1990.

81. K. Kanatani. Unbiased estimation and statistical analysis of 3-D rigid motion from two views. IEEE Transactions on Pattern Analysis and Machine Intelligence **15(1)**, 37–50, 1993.

82. P.H.S. Torr, D.W. Murray. The development and comparison of robust methods for estimating the fundamental matrix. Intern. Journal of Computer Vision **24(3)**, 271–300, 1997.

83. B. Hofmann-Wellenhof. Die gegenseitige Orientierung von zwei Strahlenbündeln bei Übereinstimmung, bei unbekannten Näherungswerten und durch ein nichtiteratives Verfahren. Doct. Thesis, Deutsche Geodät. Komm., Reihe C, **257**, München, 1979.

84. G. Monge. Darstellende Geometrie. Ostwalds Klassiker der exakten Wissenschaften Nr. 117, Engelmann Verlag Leipzig, 1900.

85. J.A. Grunert. Das Pothenotische Problem in erweiterter Gestalt nebst über seine Anwendungen in der Geodäsie. Grunerts Archiv für Mathematik und Physik **1**, 238–248, 1841.

86. K. Killian, P. Meissl. Zur Lösung geometrisch überbestimmter Probleme. Österreichische Zeitschrift f. Vermessungswesen **64(3/4)**, 81–86, 1977.

87. K. Killian. Über das Rückwärtseinschneiden im Raum. Österreichische Zeitschrift für Vermessungswesen **4**, 97–104 and **5**, 171–179, 1955.

88. A. Pope. An advantageous, alternative parameterization of rotations for analytical photogrammetry. ESSA Techn. Report Ca GS-39, Coast and Geodetic Survey, U.S. Dept. of Commerce, Rockville, Md, 1970.

89. R.M. Haralick, C. Lee, K. Ottenberg, M. Nölle. Analysis and solutions of the three point perspective pose estimation problem. Proc. IEEE Conf. on Computer Vision and Pattern Recognition, 592–598, 1991.
90. Y. Liu, Th.S. Huang, O.D. Faugeras. Determination of camera location from 2-D to 3-D line and point correspondences. IEEE Transactions on Pattern Analysis and Machine Intelligence **12(1)**, 28–37, 1990.
91. P. Lohse, W. Grafarend, B. Schaffrin. *Three-Dimensional Point Determination by Means of Combined Resection and Intersection*. Paper presented at the Conference on 3-D Measurement Techniques, Vienna, pp. 1–17, 1989.
92. E.L. Merritt. Explicit three-point resection in space. Photogrammetric Engineering **15(4)**, 649–655, 1949.
93. W. Grafarend, P. Lohse, B. Schaffrin. Dreidimensionaler Rückwärtsschnitt. Zeitschrift für Vermessungswesen **114(2–6)**, 61–67, 127–237, 172–175, 225–234, 278–287, 1989
94. E.H. Thompson. An exact linear solution of the problem of absolute orientation. Photogrammetria **15(4)**, 163, 1958–59.
95. G.H. Schut. On exact linear equations for the computation of the rotational elements of absolute orientation. Photogrammetria **15(1)**, 34–37, 1960.
96. K. Rinner, R. Burkhardt (eds.). *Handbuch der Vermessungskunde*. Bd. III, a/3 Photogrammetrie, Metzlersche Verlagsbuchhandlung, Stuttgart, 1972.
97. M. Tienstra. A method for the calculation of orthogonal transformation matrices and its application to photogrammetry and other disciplines. Publ. Intern. Institute for Aerial Survey and Earth Sciences (ITC) **A48**, Enschede, The Netherlands, 1969.
98. F. Sansó. An exact solution of the roto-translation problem. Photogrammetria **29**, 203–216, 1973.
99. B.K.P. Horn. Closed-form solution of absolute orientation using unit quaternions. Journal Optical Society of America **A4(4)**, 629–642, 1987.
100. K.S. Arun, T.S. Huang, S.D. Blostein. Least-squares fitting of two 3D point sets. IEEE Transactions on Pattern Analysis and Machine Intelligence **9(5)**, 698–700, 1987.
101. Z. Zhang, O.D. Faugeras. *3D Dynamic Scene Analysis. A Stereo Based Approach*. Springer Verlag, Berlin, Heidelberg, New York, 1992.
102. G.H. Schut. Construction of orthogonal matrices and application in analytical photogrammetry. Photogrammetria **15(4)**, 149, 1958–1959.
103. J. Stuelpnagel. On the parameterization of the three-dimensional rotation group. SIAM Review **6(4)**, 422–430, 1964.
104. S.L. Altmann. Hamilton, Rodrigues, and the quaternion scandal. Mathematics Magazine **62(3)**, 291–308, 1989.
105. K. Kanatani. Group-theoretical methods in image understanding. Springer Series in Information Sciences no. 20, Springer Verlag Berlin Heidelberg, 1990.
106. W. Kühnel, E. Grafarend. Minimal atlas of the rotation group SO(3), the set of orthonormal matrices. Report, Department of Geodesy and Geoinformatics, Stuttgart University, 1999.
107. B. Wrobel, D. Klemm. Über die Vermeidung singulärer Fälle bei der Berechnung allgemeiner räumlicher Drehungen. Intern. Archives of Photogrammetry and Remote Sensing **25**, 1153–1163, Commission III, ISPRS, Rio de Janeiro, 1984.
108. O.D. Faugeras, M. Hebert. The representation, recognition, and locating of 3-D objects. The Intern. Journal of Robotics Research **5(3)**, 27–52, 1986.

109. J. Krames. Über die "gefährlichen Raumgebiete" der Luftphotogrammetrie. Photographische Korrespondenz **84(1–2)**, 1–26, 1948.

110. O.D. Faugeras, S.J. Maybank. Motion from point matches: Multiplicity of solutions. In *Proc. IEEE Workshop on visual motion*, Irvine, Cal., pp. 248–255, 1989.

111. Th. Buchanan. Photogrammetry and projective geometry – an historical survey. SPIE – The International Society for Optical Engineering **1944**, Bellingham WA, 1993.

112. J. Krames. Über die mehrdeutigen Orientierungen zweier Sehstrahlbündel und einige Eigenschaften der orthogonalen Regelflächen zweiten Grades. Monatshefte für Mathematik und Physik **50**, 65–83, 1941.

113. J. Krames. Über die bei der Hauptaufgabe der Luftphotogrammetrie auftretenden "gefährlichen Flächen". Bildmessung und Luftbildwesen **1/2**, 1–18, 1942.

114. E. Gotthardt. Zur Unbestimmtheit des räumlichen Rückwärtseinschnittes. *Mitteilungen der Deutschen Gesellschaft für Photogrammetrie* **5**, 193–198, 1938–1940.

115. W. Wunderlich. Über den "gefährlichen" Rückwärtseinschnitt. Jahresberichte Deutsche Mathematiker Vereinigung 53, 41–48, 1943.

116. E. Gotthardt. Ein neuer gefährlicher Ort beim räumlichen Rückwärtseinschnitt. Bildmessung und Luftbildwesen **1**, 6–8, 1974.

117. E.H. Thompson. Space resection: Failure cases. Photogrammetric Record **X(27)**, 201–204, 1966.

118. Th. Buchanan. The twisted cubic and camera calibration. Computer Vision, Graphics, and Image Processing **42**, 130–132, 1988.

119. F. Schur. Über die Horopterkurve. Sitzungsberichte Naturforscher-Gesellschaft Uni Dorpat 9, 162–164, Dorpat 1889.

120. K. Killian. Der gefährliche Ort des überbestimmten räumlichen Rückwärtseinschneidens. Österreichische Zeitschrift für Vermessungswesen und Photogrammetrie **1**, 1–12, 1990.

121. B.K.P. Horn, H.M. Hilden, S. Negahdaripour. Closed-form solution of absolute orientation using orthonormal matrices. Journal Optical Society of America **A5(7)**, 1127–1135, 1988.

3 Generic Estimation Procedures for Orientation with Minimum and Redundant Information

Wolfgang Förstner

Summary

Orientation of cameras with minimum or redundant information is the first step in 3D-scene analysis. The difficulty of this task lies in the lack of generic and robust procedures for geometric reasoning, calibration and especially orientation The paper collects available tools from statistics, especially for the diagnosis of data and design and for coping with outliers using robust estimation techniques. It presents a generic strategy for data analysis on the contest of orientation procedures which may be extended towards self-calibration.

3.1 Motivation

Orientation of cameras with minimum and redundant information is the first step in 3D-scene analysis. Compared to image interpretation it looks simple, it seems to be solved in photogrammetry and is expected to be implemented within a few weeks. All experience shows that camera calibration and orientation needs much effort and the solutions provided in photogrammetric textbooks cannot be directly transferred to automatic systems for scene analysis.

The reasons for this situation lie in the hidden complexity of the calibration and orientation tasks.

- *Camera modeling* requires a thorough understanding of the physics of the image formation process and of the statistical tools for developing *and* refining mathematical models used in image analysis. High precision cameras used in aerial photogrammetry have put the burden of obtaining high precision on the manufacturer, leading to the – only partly correct – impression that calibration can be standardized, and thus is simple. The long scientific struggle photogrammetry went through in the 1970s, which is not mentioned in today's publications, must now be repeated under much more difficult boundary conditions: non-standardized video cameras, non-standardized applications, the requirement for full automation, therefore the integration of error-prone matching procedures, etc.

- The *3D-geometry of orientation* reveals high algebraic complexity. This is overseen when assuming the calibration and orientation to be known or at least approximately known, as the iterative techniques used in photogrammetry and the spatial intersection (triangulation) – in general – lead to satisfying results. Again, the efforts of photogrammetric research in the 1970s and early 1980s for generating guidelines for a good design of so-called "photogrammetric blocks", where hundreds and thousands of images are analysed simultaneously for object reconstruction, specifically mapping, has to be invested for the different tasks of 3D-scene reconstruction in computer vision, especially in the area of structure from motion. It is interesting and no accident that such guidelines are only available for aerial photogrammetric blocks, not for close range applications. The complexity of the 3D-geometry of orientation motivated the numerous publications in the computer vision area on the availability, uniqueness and stability of orientation and reconstruction procedures under various, partly very specific, boundary conditions.

- *Error handling* is a central issue in calibration and orientation of cameras for several reasons.

  - The *correspondence problem* is far from being solved for general cases. Existing solutions have to deal with large percentages of matching errors. This prevents the direct use of classical estimation procedures and makes it necessary to look for robust procedures which, however, make a thorough analysis of the quality of the final result at least difficult, as the underlying theories (!) often only give asymptotic theorems.

  - In case *approximate values* for calibration and orientation are not available or are only of poor quality their determination appears to be a far more challenging problem than the refinement via a least squares estimation. The direct solutions, either with minimum or redundant information, play a central role, especially in the presence of outliers.

  - *Self-calibration* is often required where calibration, orientation and generally also scene reconstruction is performed simultaneously, as camera calibration in a laboratory is often not feasible or is insufficient. It increases the difficulty of error analysis by at least one order of magnitude as deficiencies in design, modelling and mensuration have to be handled simultaneously and, therefore, generally prevent an algebraic analysis of the system. The difficulty of integrating *all* types of observational values lies in the necessity to formalize the evaluation process in order to adequately handle the different dimensions (pixels, meter, radiants, etc.) of the observations and their influence on the final result.

Experience in photogrammetric research gives many hints on how to solve the problem of error handling, especially with respect to the quality evaluation based on various statistical tools. Nonetheless, the boundary conditions met in computer vision applications require a new setup of the concepts.

- The final goal of image analysis is *full automation* of all procedures. As calibration and orientation of cameras, due to its well-defined goal, really is much simpler than image interpretation, it seems to be feasible to achieve generic procedures for automatically solving this first step in the analysis chain. Textbooks on photogrammetry, statistical analysis or other related topics, however, often only *present tools not strategies* for solving the problem of parameter estimation, calibration and orientation like many other subtasks in image analysis. This is due to the specific engineering expertise which is required to find the appropriate tool combination. This expertise is usually not documented in textbooks, but in internal reports of institutions for training purposes, e.g. for handling complex software packages. Sometimes this knowledge is already formalized in terms of a sequence of rules to be applied.

  Formalization, being a prerequisite for developing generic procedures, is difficult in our context as the various types of errors (cf. Sect. 3.2.2 on error handling) interfere in a nonpredictable manner and no coherent theory is available to justify specific strategies.

This chapter is motivated by this deficit in generic and robust procedures for geometric reasoning, calibration and especially orientation. Its aim is to collect the available tools from statistics, specifically for the diagnosis of data and design and for coping with outliers using robust estimation techniques, and to present a generic strategy for data analysis in the context of orientation procedures. The techniques allow an extension towards self-calibration which, however, has to be worked out. The much more difficult problem of designing, i.e. planning mensuration procedure of high robustness, still waits for a solution.

3.2   Problem Statement

Let us assume the model to explicitly describe the observation process

$$E(\boldsymbol{b}) = \boldsymbol{g}(\boldsymbol{x}) \tag{3.1}$$

where the expectation of the n observations $\boldsymbol{b} = \{b_i\}$ via $\boldsymbol{g}$ in general nonlinearly depends on the u unknown parameters $\boldsymbol{x} = \{x_j\}$. The stochastic properties of the observations are captured by the covariance matrix

$$D(\boldsymbol{b}) = \boldsymbol{C}_{bb}. \tag{3.2}$$

Should this be the only information available the principle of maximum entropy results in the following full model

$$b \sim N(g(x), C_{bb}) \tag{3.3}$$

hypothesizing b to be normally distributed. The redundancy of the system is

$$r = n - u. \tag{3.4}$$

The task is to derive estimates $\hat{x}$ from given observational values b.

In our context the observations are usually the coordinates of points or the parameters of lines detected and located in the image by an automatic procedure. The relation between corresponding points and/or lines in several images or in object space, also performed automatically, guarantees redundancy in the total process, as several image features generally determine one corresponding object feature.

In case the redundancy equals 0 or in the unlikely case of the observations being consistent, the assumed stochastic properties have no influence on the estimate. The only task then is to invert (3.1) to obtain $\hat{x} = g^{-1}(b_s)$, where b_s is a subset of b of size u.

3.2.1   Error Types

In general, all components of the model will have an influence on the result. The key question is how an automatic system handles errors in these assumptions. One may distinguish three types of errors:

1. *Data errors*, which are errors in the values of b, grossly violate assumption (3). They relate to points, lines or other features in the image or in object space where measurements are taken. They may really be mensuration errors, e.g. caused by failures in the detection algorithm or matching errors leading to wrong relations between image and object features. Depending on the complexity of the scene and the quality of the used algorithms the percentage of errors may range between a few and over 80% of the observed values.

2. *Model errors* refer to all three parts of the model: the functional relationship $g(x)$, the covariance matrix C_{bb} and the type of the distribution, here the normal distribution $N(\cdot, \cdot)$. Examples for this type of error are manifold:
   - too few, too many or the wrong set of parameters x, e.g. when using shallow perspective, projectivity or parallel projection;
   - wrong weighting, e.g. when assuming the same accuracy for all detected points;
   - neglected correlations, e.g. in Kalman filtering; or,
   - wrong assumptions about the distribution, e.g. when handling one-sided errors.

Observe that data errors and model errors cannot formally be distinguished; as a refinement of the model one may always specify the type of error in the observations.

3. *Design or configuration errors* relate to the complete set of functions $g = \{g_i\}$. Such errors cause the estimate $\hat{x}$ to be nonunique in some way. Multiplicity of solutions is the best case of nonuniqueness. Depending on the degree of redundancy we may distinguish at least three cases (cf. the formalization in Sect. 3.3.2):

   (a) nondeterminable parameters. Critical surfaces of the configuration belong to this class. An example would be a spatial resection with three points and the projection center sitting on the critical cylinder.

   (b) noncheckable observations or parameters. Here the determination of the parameters may be possible, but errors in the estimated parameters introduced in a Bayesian manner are not detectable due to a too low redundancy. An example would be a spatial resection with three points in general position.

   (c) nonlocatable errors. Here a test may be able to show discrepancies between the data and the model, but no identification of the error source is possible. An example would be a spatial resection with four points in general position.

We will treat all types of errors in the following; however, we will concentrate on means for automatically reacting on indications of such errors.

3.2.2 Issues in Error Handling

There are at least three basic questions that automatic procedures need to be able to answer:

1. How sensitive are the results?
   The results may be uncertain due to the large number of errors mentioned above. Evaluating real cases has to cope with the problem that several such errors occur simultaneously. *Instabilities* due to low local redundancy may be hidden within a system of high total redundancy. Then we may discuss

   • determinability of parameters
   • controllability of errors and the effect of nondetectable errors
   • separability of error sources.

   We will formalize this classification in more detail and discuss the first two items explicitly.

2. How small is too small?
   Most algorithms are controlled by *thresholds* or tolerances to be specified by the developer or the user.
   When referring to observations or parameters, thresholding may be interpreted as hypothesis testing, which allows one to derive the thresholds

by specifying a significance level and using error propagation. We will not pursue this topic.

When evaluating, the *design* of the formalization becomes less obvious, e.g. when having a small basis in relative orientation (2D–2D), small angles in spatial resection (3D–2D) or small distances between all points in absolute orientation (3D–3D). In all cases the configuration is close to critical. But then the question arises: how to evaluate *small deviations from a critical configuration or surface?* We will show that a generic and formal answer to this question can be given which is based on the *local* geometry of the design.

3. How to react to deficiencies in the data?

Regarding the many different models used for calibration and orientation a *generic strategy* should be available.

Deficiencies in design have to be prevented by proper planning of the mensuration setup influencing the number and position of cameras, the number and the distribution of given control points, the introduction of spatial constraints, etc. Automized techniques for such planning are not far advanced and still require interactive intervention.

The reaction on *deficiencies in the observations* or the model may rely on the techniques from robust estimation and much more from formalizable experience.

They depend on various properties of the data and the model:

- the availability of approximate values x^0 for the unknown parameters x.
- the availability of a direct solution $x = g^{-1}(b_s)$ for a u-sized subset of the observations.
- the number and the size of the expected errors.
- the number of the observations and parameters
- the desired efficiency in terms of computing time, etc.

The next section will collect the necessary tools needed for setting up generic procedures for robust estimation applicable to camera orientation.

3.3 Tools

3.3.1 Quality Assurance

Treating calibration and orientation as an estimation problem allows us to fully exploit the rich arsenal of tools from estimation theory. Regarding the specific problem of data and model errors we specifically need to use the techniques available from robust statistics and regression diagnostics following two different aims ([13]):

- The purpose of *robustness* is to have safeguards against deviations from the assumptions.
- The purpose of *diagnostics* is to find and identify deviations from the assumptions.

Robustness. There are two levels of robustness, depending on whether the size of errors is small or large. Data or model deviations are small in the case of sufficient linear approximations. This leads to a rule of thumb that small deviations of the approximate values from the true values are deviations less than about 30% of the values, including all functions of the observations. E.g., it corresponds to the requirement that angular errors be less than approximately 20°.

1. Robustness with respect to *small deviations.*
   The so-called *influence curve* [11], which measures the effect of errors on the result, may be used to measure the quality of robust procedures in this case. Maximum-likelihood (ML) type, or *M-estimators* are the tools to deal with small deviations.
2. Robustness with respect to *large deviations.*
   The *break down point* [16] measuring the maximum allowable percentage in the number of errors while still guaranteeing the estimator to yield results with limited bias, may be used to evaluate the quality of procedures in this case. Estimates with a high break down point, up to 50%, such as least median squares, are the corresponding tool to handle a large percentage of errors.

Observe that the effect of *random* errors on the result is not covered by the term "robustness". These effects are usually measured by the *precision* of the estimates. The reason for this distinction is that random errors are part of the original model, thus do not represent deviations from the model, and are taken into account by all basic estimators such as least-squares or ML-estimators.

We will discuss the use of different robust estimators in Sects. 3.3.4 and 3.4, where we especially compare and use their characteristics for achieving a generic strategy.

Diagnostics. As already indicated above, there are three levels of diagnostics which all refer to small model errors:

1. *Determinability* of parameters or singularities in the estimation process measure the instability of the design with respect to random perturbations.
   Standard deviations or in general covariance matrices are the diagnostic tool to detect such a situation. Due to the small size of the random errors, a linear substitute model derived by linearization may be used to evaluate such instabilities.
   We will discuss this in detail in Sect. 3.3.2.
2. *Controllability* of observations and detectability of model errors specify the applicability of hypothesis tests.
   The diagnostic tools are minimum bounds of the size of observational or model errors which can be detected by a test with a certain given

probability. The *sensitivity* of the result is measured by the effect of non-detectable errors on the result.

Both tools may be used for planning as they do not depend on the actual measurements.

The actual influence of the observations of model parameters measured in a leave-one-out fashion may be decisive for the acceptance of an estimate. We will discuss these tools in detail in Sect. 3.3.3.

3. The *locatability* of observational errors or the separability of model errors specify the ability to correctly classify or identify the error causes.

This can be described in terms of a confusion matrix, like in statistical pattern recognition, the difference being that here the entries of the confusion matrix depend on the expected size of the errors and on the design or configuration.

The diagnostic tools therefore are lower bounds for observational errors or model errors which are identifiable or separable with a certain probability. In Sect. 3.3.3 we will formally relate separability to controllability especially with respect to *sets* of observational model errors, but will not discuss the notion in detail.

3.3.2   Instabilities of Estimates or "How Small is too Small?"

Instabilities of parameters occur in case the configuration produces some critical manifold (surface) to which the solution belongs. One usually distinguishes (cf. [20]):

1. Singularities or critical surfaces of the first kind. Here a complete manifold of the parameters is consistent with the observations.
2. Singularities or critical surfaces of the second kind. Here small deviations in the observations result in large deviations in the parameters.

An example for a singularity of the second kind is the critical cylinder in spatial resection. It may be formulated as a rule: IF the projection center $O \in$ cylinder(P_1, P_2, P_3) THEN O is not determinable. Here cylinder (P_1, P_2, P_3) indicates the cylinder through the points with axis perpendicular to the plane through the points.

This rule is the result of an analysis using algebraic geometry which, in its generality, is valid in the context of spatial resection and is crisp.

Such algebraic results, however, have some disadvanteges:

- The statements do not contain any information on how to evaluate *deviations from the critical configuration*.
- The statements do not give any hint to *generalize* to other situations. Other problems, e.g. relative orientation, require a separate analysis.
- The statements do not give any *means to evaluate* the orientation even of one image *within a set* of several images to be oriented simultaneously. It may very well be that in a multi-image setup with a large redundancy

the orientation of one of the images cannot be determined due to the presence of the above situation.

Such *hidden instabilities* reveal the limitation of purely algebraic approaches which can only be applied to very restricted situations and cannot be generalized.

Thus techniques based on algebraic geometry cannot be easily transferred into automatic procedures evaluating the stability of an estimate. The solution to this dilemma is based on the observation, that the instabilities are local properties in parameter space and can be fully analysed using the covariance matrix of the parameters. This leads to a shift of the problem. Instead of a deterministic analysis we are now confronted with the problem of evaluating the quality of a covariance matrix. The shift of the problem and its solution goes back to Baarda [1].

The evaluation method consists of two steps:

1. Specification
   Specifying the user requirements in terms of a so-called *criterion matrix*, say H, which gives an upper bound on the desired covariance matrix, corresponding to the desired lowest precision.
2. Comparison
   Checking whether the achieved covariance matrix, say $G = (A^T C_{bb} A)^{-1}$ is better than H.

We will discuss this comparison first.

Comparing Covariance Matrices. The *comparison* of covariance matrices is interpreted as the requirement that the standard deviation of an arbitrary function f should be better when calculated with covariance matrix G than with H

$$G \leq H \quad \doteq \quad \sigma_f^G \leq \sigma_f^H, \text{ with } f = e^T \widehat{x}, \text{ for all } e \qquad (3.5)$$

Using error propagation, e.g. $\sigma_f^G = \sqrt{e^T G e}$ this leads to (cf. Fig. 3.1)

$$e^T G e \leq e^T H e, \text{ for all } e \qquad (3.6)$$

or

$$\lambda = \frac{e^T G e}{e^T H e} \leq 1 \qquad (3.7)$$

which requires the determination of the maximum eigenvalue of

$$G e = \lambda H e. \qquad (3.8)$$

The square root $\sqrt{\lambda_{max}}$ indicates the maximum ratio of the actual and the required standard deviation.

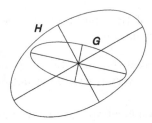

Fig. 3.1. The relation $G < H$ between two 2×2 covariance matrices G and H, represented by isolines of constant probability density of the corresponding normal distribution.

This evaluation may be simplified using

$$K = H^{-1/2}GH^{-1/2} \qquad (3.9)$$

$$\lambda = \frac{e^{\mathrm{T}}Ke}{e^{\mathrm{T}}e} \le 1 \qquad (3.10)$$

which is equivalent to

$$\lambda_{\max}(K) \le 1. \qquad (3.11)$$

Equation (3.10) is favorable in case H can easily be diagonalized (cf. the example below).

In order to avoid the rigorous determination of the maximum eigenvalue of K, (3.10) may be replaced by a less tight norm, e.g. by the trace:

$$\lambda_{\max}(K) \le \mathrm{tr}K \le 1. \qquad (3.12)$$

Specification of a Criterion Matrix. The *specification* of a criterion matrix can be based on the covariance matrix $C_{\widehat{x}\widehat{x}}$ derived from an ideal configuration. This has the advantage that the user can easily interpret the result. In case an ideal configuration cannot be given the criterion matrix $H = SRS$ may be set up by specifying the standard deviations σ_i, collected in a matrix $S = Diag(\sigma_i)$ and correlations ρ_{ij}, collected in a matrix $R = \rho_{ij}$, derived from some theoretical considerations, e.g. interpreting the sequence of projection centers in a navigation problem as a stochastic process, where the correlations ρ_{ij} depend only on the time or space difference between points P_i and P_j.

Example. Five image points situated as in Fig. 3.2 are to be used to estimate the six orientation parameters of the image based on given 3D coordinates with spatial resection (2D–3D). Due to gross errors in the data, a RANSAC procedure (cf. [4]) is applied randomly selecting three points and directly solving for the orientation parameters. The quality of this selection has to

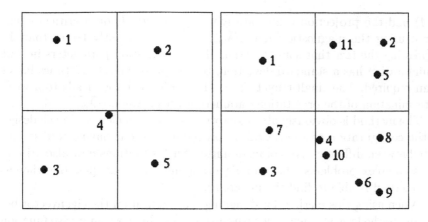

Fig. 3.2. Two sets of image points used for image orientation by spatial resection. Sets of three points may lead to results of different stability as shown in Table 3.1 for three sets of the left configuration (from Schickler [17]).

be evaluated automatically in order to immediately exclude unstable configurations. The above mentioned technique for evaluating the stability of a configuration is applied.

The criterion matrix is derived from a very stable least-squares fit with four points symmetrically sitting in the four corners of the image (cf. Appendix). The covariance matrix $C = C_{\widetilde{xx}}$ of this configuration, the criterion matrix, is chosen to be

$$H = 16 \cdot C \qquad (3.13)$$

thus requiring the standard deviations of the orientation parameters within the RANSAC-procedure to be better than four times the standard deviation of the ideal configuration. C is sparse allowing us easily, i.e. algebraically, to determine the matrix $H^{-\frac{1}{2}}$ in (3.10) (cf. Appendix).

For several triplets of points the ratio $\sqrt{\lambda_{\max}}$ is given.

Table 3.1. The stability with sets of three points used for spatial resection (cf. Fig. 3.2a).

	Configuration	$\sqrt{\lambda_{\max}}$
1	1/2/3	0.8
2	2/3/4	88.0
3	1/3/4	13.2

The good triangle (1, 2, 3) obviously leads to sufficiently precise orientation parameters. The second triplet (2, 3, 4) consists of three nearly collinear points, which is obviously an undesirable configuration. The third triplet (1,

3, 4) and the projection center are lying *approximately* on a critical cylinder causing the diagnostic value $\sqrt{\lambda_{\max}}$ to be significantly larger than 1., expressing the fact that some function of the orientation parameters in that configuration has a standard deviation, being approximately 13 times larger than required. The small triplet (2, 5, 11) in Fig. 3.2b also leads to a weak determination of the orientation parameters with a value $\sqrt{\lambda_{\max}} \approx 4$.

The method is obviously able to capture various deficiencies in the design of the configuration of an orientation procedure without having to discriminate between different types of instabilities. Such situations may also arise in more complex problems where an algebraic analysis is not possible whereas this method is able to find the instabilities.

When using this method for designing a configuration the eigenvector belonging to the largest eigenvalue gives insight into the most imprecise function of the parameters, which may be used to look for specific stabilization means.

3.3.3   Model Errors or "How Sensitive is the Result?"

The stability of an estimation, specifically an orientation, evaluated by the covariance matrix only takes random perturbations into account. The result, however, may be wrong due to gross errors, e.g. caused by the matching procedure. In addition, an oversimplified model may lead to precise but incorrect results. Both error sources, blunders and systematic errors, can only be detected in the case of redundant observations. This is a necessary but – as we will see – not a sufficient condition. Redundancy allows us to perform tests on the validity of the assumed model *without* reference to additional data used during the estimation. Such tests may lead to the detection or even identification of the error source. Of course, the outcome of these tests may be false. Redundancy, however, increases the stability of the solution and the correctness of the outcome of statistical tests. The theory for performing such a test is described in the literature (cf. [2,3,6,8]). The structure of that theory, its use in estimation problems and examples from orientation procedures will be given.

Detectability and Separability of Errors. We first want to discuss the type of evaluation which can be performed depending on the redundancy r of a system.

1. $r = 0$ In the case of no redundancy, one can only evaluate the sensitivity of the result with respect to random errors as shown in the last section. No check of the observations is possible whatsoever. They may remain incorrect without any indication.
2. $r = 1$ In the case of redundancy $r = 1$, a check on the validity of the model is possible. The existence of blunders may be indicated. However, they are not locatable, as a "leave-one-out test" always leads to a valid solution.

3. $r = 2$ A redundancy of $r = 2$ is necessary in order to be able to locate a single blunder. A leave-one-out test generally will be able to find the unique consistent set of observations. Double errors are not locatable, however their existence is usually indicated.
4. $r > 2$ For a larger redundancy, $r - \cdot 1 < n/2$, errors are locatable, whereas r errors are only detectable.

The maximum number of detectable errors is $n/2$, i.e. 50% of the data, as more than $n/2$ observations may mimic a good result. Thus, 50% is the upper limit for the so-called *breakdown point* of an estimator. The breakdown point of an estimator is the minimum percentage of errors which may cause the estimator to give wrong results, i.e. may lead to a bias of any size. The normal mean has the breakdown point 0, the median 50%, an indication of its higher robustness. Practical procedures may be better as they may use specific knowledge about the structure of the problem (cf. the straight line detection procedure by Roth and Levine [15]).

In case of a small percentage ($< 1\%$) of not too large ($\leq 30\%$) gross errors, the detection and location may be based on the residuals

$$v = g(\widehat{x}) - b \qquad D(l) = \sigma_0^2 Q = \sigma_0^2 P^{-1} . \tag{3.14}$$

Using the maximum likelihood estimate

$$\widehat{x} = x^{(0)} + (A^T P A)^{-1} A^T P(l - g(x^{(0)})) \tag{3.15}$$

we can express changes Δv of the residuals in terms of changes, thus errors Δl of the observations

$$\Delta v = -R \Delta l \tag{3.16}$$

with the projection matrix
$$R = I - U \tag{3.17}$$

with the so-called hat-matrix [14]
$$U = A(A^T P A)^{-1} A^T P . \tag{3.18}$$

(3.17) is graphically shown in Fig. 3.3.

This matrix may be used to analyse the ability of the estimation system to apply self-diagnosis with respect to errors in the observations, as only effects that can be seen in the residuals are detectable.

We distinguish two levels of evaluation:

1. detectability or checkability; and,
2. separability or locatability.

Both evaluation measures may refer to single or groups of observations. Thus we have four cases.

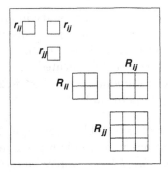

Fig. 3.3. The four cases for analysing the projection matrix **R** with respect to sensitivity (diagonal matrices) and separability (off-diagonal matrices) for single or groups of observations.

1. *Detectability* or checkability rely on the diagonal elements or diagonal submatrices of $\boldsymbol{R}$.

   a) *Single* observational errors can only be detected if the *redundancy numbers*

   $$r_i \doteq (\boldsymbol{R})_{ii} > 0. \tag{3.19}$$

   The diagonal elements r_i sum up to the total redundancy r, i.e. $\sum r_i = r$. This indicates how the redundancy is distributed over the observations. The corresponding test statistics for detecting single errors for given σ_0 and uncorrelated observations is

   $$z_i = \frac{-v_i}{\sigma_0} \sqrt{\frac{p_i}{r_i}} \sim N(0,\,1). \tag{3.20}$$

   b) *Groups* of n_i observation can only be detected if the corresponding $n_i \times n_i$ submatrix

   $$\| R_{ii} \| > 0 \tag{3.21}$$

   of $\boldsymbol{R}$ is nonsingular. Otherwise a special combination of observational errors may have no influence on the residuals. The corresponding test statistic is

   $$T_i = \frac{1}{\sigma_0} \sqrt{\frac{\boldsymbol{v}_i^{\mathrm{T}} \boldsymbol{R}_{ii} \boldsymbol{Q}_{ii} \boldsymbol{v}_i}{n_i}} \sim \sqrt{\boldsymbol{F}_{n_i,\infty}}\,, \tag{3.22}$$

   which reduces to (3.20). The observations may be correlated within the group, but must be uncorrelated to the others. $\sqrt{\boldsymbol{F}_{n_i,m}}$ denotes the distribution of the square root of a random variable being $F_{n_i,m}$-distributed.

2. *Separability* or locatability in addition to the diagonal elements of $\boldsymbol{R}$ rely on the off diagonals.

a) The separability of two *single gross errors* evaluates the likelihood to correctly locate an error, i.e. to make a correct decision when testing both. The decisive measure is the correlation coefficient of the test statistics (3.20) which is

$$\rho_{ij} = \frac{r_{ij}}{\sqrt{r_{ii} \cdot r_{jj}}}. \tag{3.23}$$

Tables for erroneous decisions when locating errors are given by Förstner [9].

Correlation coefficients below 0.9 can be accepted since the probability of making a false decision even for small errors remains below 15%.*

b) The separability of two *groups of observations* $\boldsymbol{b}_i$ and $\boldsymbol{b}_j$ depends on the maximum value

$$\rho_{ij}^2 = \lambda_{\max} \boldsymbol{M}_{ij} \tag{3.24}$$

of the $n_i \times n_j$ matrix

$$\boldsymbol{M}_{ij} = \boldsymbol{R}_{ij} \boldsymbol{R}_{jj}^{-1} \boldsymbol{R}_{ji} \boldsymbol{R}_{ii}^{-1} \tag{3.25}$$

which for single observations reduces to (3.23).

No statistical interpretation is available due to the complexity of the corresponding distribution.

Example: *Detectability of Errors*

Relative orientation with six corresponding points yields a redundancy of $r = 6 - 5 = 1$. If the images are parallel to the basis and the points are situated symmetrically as shown in Fig. 3.4 then the diagonal elements r_i are $1/12$ for points $i = 1, 2, 5$ and 6 and $1/3$ for points 3 and 4.

Obviously errors are hardly detectable if they occur in point pairs 1, 2, 5 or 6. In all cases no location of the false matches is possible as $r = 1$. □

Example: *Separability of Errors*

Spatial resection with four points symmetrically to the principle point is known to yield highly correlated orientation parameters. Depending on the viewing angle α, the correlation between the rotation ω (x-axis) and the coordinate y_0 of the projection center, and between the rotation φ (y-axis) and the coordinate x_0 is (cf. Appendix)

---

* Precisely stated: If the larger of the two test statistics $|z_i|$ and $|z_j|$ in (3.20) is chosen to indicate the erroneous observation with its critical value 3.29, corresponding to a significance level of 99.9%, and a single error can be detected with a probability higher than 80%, then the probability of making a wrong decision between $\boldsymbol{b}_i$ and $\boldsymbol{b}_j$ is approximately 13%.

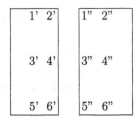

Fig. 3.4. Numbering of six points in a stereo pair.

$$| \, \rho \, | = \frac{1}{\sqrt{1 + \sin^4 \frac{\alpha}{2}}} \, . \qquad (3.26)$$

For a CCD camera with a focal length of $f = 50$ mm and sensor size of $5 \times 5\,\text{mm}^2$, $\alpha/2 = 1/20$ thus $| \, \rho \, | = 0.999997$. For an aerial camera RMK 15/23 with a focal length of 15 cm and image size of 23 cm, $\alpha/2 = 2/3$, thus $| \, \rho \, | = 0.914$.

Thus testing the orientation parameters ω, φ, x_0 and y_0 may easily lead to incorrect decisions for CCD cameras when testing their significance, whereas errors in these parameters are detectable. $\square$

Sensitivity of the Estimates. In spite of testing for blunders, errors may remain undetected and influence the resulting estimate. The *sensitivity* of the result is often the only information one needs for evaluation. One may determine an upper limit for the influence of a group of observations on the result.

The influence $\Delta_i f(\widehat{x})$ on a function $f(\widehat{x})$ of the unknown parameters caused by leaving out a group l_i of observation is limited:

$$\Delta_i f(\widehat{x}) \leq \Delta_i f_{\max}(\widehat{x}) \qquad (3.27)$$

with (cf. [10])

$$\Delta_i f_{\max}(\widehat{x}) = T_i \cdot \mu_i \cdot \sigma_{f(x)} \cdot \sqrt{n_i} \qquad (3.28)$$

where n_i is the size of the group; $\sigma_{f(x)}$, the standard deviation of the function $f(x)$, is derivable by error propagation measuring the precision of the result, T_i of the test statistics (3.22), measuring the quality of the observation group and the geometry factor

$$\mu_i = \lambda_{\max}\{(C_{xx}^{(i)} - C_{xx})C_{xx}^{-1}\} \qquad (3.29)$$

evaluating the mensuration design. The value μ_i explicitly measures the loss in precision, i.e. the normalized increase $C_{xx}^{(i)} - C_{xx}$ of the variance of the result when leaving out the ith group b_i of observations.

For a single observation it reduces to

$$\mu_i = \frac{1 - r_i}{r_i} \tag{3.30}$$

with the diagonal elements r_{ii} of $\boldsymbol{R}$ (cf. (3.17)).

The value $\Delta_i f_{\max}(\boldsymbol{x})$ (3.28) measures the *empirical sensitivity* of the estimate with respect to blunders e.g. matching errors in groups $\boldsymbol{b}_i$; empirical, as it depends on the actual observations via T_i.

If T_i is replaced by a constant δ_0, indicating the minimum detectable (normalized) error, we obtain the *theoretical sensitivity*

$$\Delta_{0i} f(\widehat{\boldsymbol{x}}) \leq \Delta_{oi} f_{\max}(\widehat{\boldsymbol{x}}) \tag{3.31}$$

with

$$\Delta_{0i} f_{\max}(\widehat{\boldsymbol{x}}) = \delta_0 \cdot \mu_i \cdot \sigma_{f(x)} \cdot \sqrt{n_i} \,. \tag{3.32}$$

It may be used for planning purposes since it does not depend on actual observations and can therefore be determined in advance. δ_0 is usually chosen to be larger than the critical value k for T_i, e.g. $\delta_0 = 1.5k$ or $\delta_0 = 2k$ and can be linked to the required power of the test (cf. [2,3,8]).

Observe that both sensitivity values contain the product of terms representing different causes. This e.g. allows us to sacrifice precision, thus increasing the standard deviation σ_f by paying more for leaving a larger redundancy and lowering the geometric factor μ_i for all observations or vice versa.

Example: *Sensitivity Analysis*

This example shows the power of this type of sensitivity analysis for evaluating the success of an automatic procedure for determining the exterior orientation of an image, i.e. the extrinsic parameters of the camera (SCHICKLER [17], cf. [18]). It is based on matching 2D line segments in the image with 3D line segments of a set of known objects, mainly being buildings represented by a set of line segments. The aerial images used here usually contain 5–10 such objects which are more or less well distributed over the field of view.

The sensitivity analysis may be used to evaluate the quality of the orientation with respect to

(a) matching errors of *individual line segments*; and,
(b) matching errors of complete *sets* of line segments, representing one *object*.

The reason for this distinction is that both errors may occur; the first one being very common, the second one (whole sets of line segments) within the clustering procedure performed for each object individually.

(a) Matching of *individual* 2D image line segments to 3D object line segments. We have to deal with groups of four observations, namely the four coordinates representing the start and end points of the line segments. The 4×4 covariance matrix $C_{l_i l_i}$ of this group also contains the correlations between the coordinates, which may be derived during the edge extraction process. We use a similar approach as Deriche and Faugeras [7] and Förstner [10] for representing the uncertainty of the line segments.

A typical result, as given in Table 3.2, can be summarized in two statements:

1. Empirical sensitivity: The maximum occurs at edge #10. The result may change up to 0.82 times its standard deviation if line segment #10 would be left out, which is fully acceptable.
2. Theoretical sensitivity: The maximum occurs at edge #21. The result may change up to 4.42 times its standard deviation if a matching error remains undetected, which is at the limit of being acceptable.

Thus, the result appears to be acceptable with respect to the redundancy in the estimations.

Table 3.2. The empirical and theoretical sensitivity of the result of an orientation with straight edge segments.

	Empirical	Theoretical
Edge #	$\Delta_i f / \sigma_f$	$\Delta_{0i} f / \sigma_f$
4	0.07	2.62
5	0.65	1.51
8	0.50	3.44
9	0.80	3.13
10	0.82	2.81
⋮	⋮	⋮
21	0.68	4.42
⋮	⋮	⋮

(b) Match of a *set* of 2D image line segments to 3D object line segments. Let us assume the m sets of segments to be matched, have $k_i, i = 1, \cdots, m$ line segments each, and we have to fear a matching error for a complete set. Then the sensitivity analysis has to be based on sets of $4 \times m_i$ coordinates for the m_i line segments.

Figures 3.5a, b show the position of the sets within two aerial images ($c = 15$ cm) to be oriented.

In Fig. 3.5a, one of the five sets, namely #3, was not matchable, leaving the spatial resection with four objects in the three other corners and in the middle of the image. The circles around these "four points" have a radius proportional to $\delta_{0i} = \Delta_{0i} f_{\max} / \sigma_f$ and indicate how sensitive the

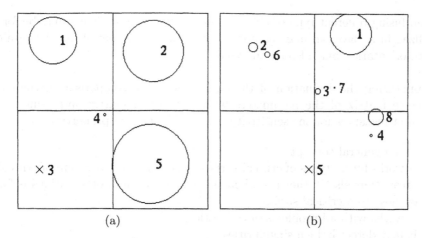

Fig. 3.5. Two sets of image points used for image orientation by spatial resection. The radii of the circles indicate the theoretical sensitivity, i.e. the amount the result might change if the point (set of straight line segments) is wrong without being noticed. In (a) the point #3 has been detected to be wrong, thus only four points are left for spatial resection; in (b) point # 5 has been detected to be wrong.

orientation is with respect to nondetectable errors within the clustering procedure. Because the geometry factor μ (3.29) is dominant, the circles indicate how the precision deteriorates if one of the four sets is left out:

without set 4: the three others 1, 2 and 5 form a well-shaped triangle, and thus guarantee good precision.

without set 2: the three others 1, 4 and 5 nearly sit on a straight line leading to a highly unstable solution (near to a singularity of first type).

without set 1: the three others, 2, 4 and 5, form a well-shaped triangle. However, because the plane going through the sets is nearly parallel to the image plane, the projection center lies close to the *critical cylinder*.

leaving out set 5: also leads to a nearly singular situation.

The situation with eight sets in Fig. 3.5b shows a more irregular distribution. Since set 5 was not matched, set 1 is most influential in the orientation, but less than sets 1, 2 and 5 in the case of Fig. 3.5a.
□

Observe that this analysis is based on values which have a very precise geometric meaning. This allows for an easy definition of thresholds, even if one is not aquainted with the underlying theory. As well, a clear comparison between different configurations is possible even for different types of tasks. Because the evaluation refers to the final parameters, it may also be used when fusing different type of observations. As model knowledge may be formalized in a Bayesian manner, the effect of prior information on the result of an orientation may also be analysed.

Summarizing the evaluation of the design using the comparison of the covariance matrix of the parameters with a criterion matrix and using the different measures for the sensitivity has several distinct properties:

- it is a general concept
- it works for all types of critical surfaces and solves the problem of critical areas, thus also in the case when the configuration of observations is far or close to a critical surface
- it works with all problems of estimation
- it may detect hidden singularities
- it also works in the complex situation where observations of different types are mixed (points, lines, circles, ...) or in the context of sensor fusion where also physical measurements (force, acceleration, ...) are involved
- it is related to a task, thus explicitly depends on user requirements. This enables us to argue backwards and optimize the design.
- it provides measures which are easily interpretable.

3.3.4   Robust Estimation or "How to React on Blunders"

The last section clearly demonstrated that enough tools are available to evaluate the result of estimation procedures with respect to a variety of deficiencies. These tools are sufficient for proving a result to be acceptable. They, however, give no hint as to how to reach an acceptable result with respect to errors in the data and weaknesses in the design.

This section wants to collect the techniques from robust statistics useful for the efficient elimination or compensation of outliers in the data with the aim of adapting the data to the presumed model. The planning of the mensuration design is much more difficult and lacks enough theoretical basis and is therefore not discussed here.

Eliminating blunders is a difficult problem:

- It is NP-complete: given n observations there are up to 2^n hypotheses for sets of good and bad values (the power set of n observations), making an exhaustive search for the optimized solution obsolete except for problems with few observations.
- The non-linearity of most estimation problems, particularly orientation problems, prevents generic simplification for obtaining suboptimal solutions.

- All variations of "Murphy's Law" occur:
  - outliers cluster and support each other,
  - outliers mimic good results,
  - outliers hide behind configuration defects,
  - outliers do not show their causes, making proper modeling difficult or impossible,
  - outliers make themselves indistinguishable from other deficiencies in the model, like systematic errors.

Thus many methods for robust estimation have been developed. Most of them assume the model of a mixed distribution of the residuals v_i (f denoting a density function here):

$$f(v_i/\sigma_i) = (1 - \varepsilon)\phi(v_i/\sigma_i) + \varepsilon h(v_i/\sigma_i) \qquad (3.33)$$

with $100\varepsilon\%$ outliers having a broad distribution $h(x)$ and $100(1 - \varepsilon)\%$ good observations following a well-behaved distribution ϕ, usually a Gaussian. Maximizing $f(x \mid b)$ or minimizing $-\log f(x \mid b)$ for the given data b, possibly including prior knowledge of the unknowns x, explicitly or implicitly is used as the optimization criterion.

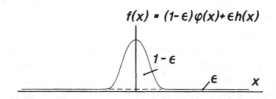

Fig. 3.6. The distribution of the errors can be a mixture of good and bad ones.

The procedures, however, significantly differ in strategy for finding the optimum or a suboptimum. We selected four procedures which seem to be representative in order to come to an evaluation which will be the basis for the generic strategy discussed in Sect. 3.4. These techniques for robust estimation are

1. complete search,
2. Random Sample Consensus (RANSAC cf. [4]),
3. clustering, and
4. ML-type estimation [14,11].

Their feasibility and efficiency heavily depend on a number of characteristic features of the estimation problem to be solved.

(a) *Invertibility of the Functional Model*
We basically used a set b_s of at least u observation to uniquely specify the unknown parameters x. The direct determination of x from a subset b_s requires g to be invertible: $x(b_s) = g^{-1}(b_s)$ thus g^{-1} has to be representable algebraically.

(b) *Existence and Quality of Approximate Values*
In case $g(b)$ is not invertible, we need approximate values for x in order to solve $x = g^{-1}(b)$ by *some* iterative scheme. The quality of the approximate values directly influences the number of iterations. The knowledge of good approximate values in all cases may drastically reduce the complexity of the procedures.

(c) *Percentage of Gross Errors*
The percentage of gross errors may range from $< 1\%$, specifically in large data sets derived automatically, up to more than 75%, e.g. in matching problems. Not all procedures can cope with any percentage of errors, some are especially suited for problems with high outlier percentages. ML-type estimation procedures can handle data with a moderate number of errors, up to 10–20% say.

(d) *Size of Gross Errors*
Only few procedures can work for any size of gross errors. Large gross errors may lead to leverage points, i.e. to locally weak geometry, and such errors may not be detectable at all. If one relates the size of the errors to the size of the observed value, then errors less than one unit are usually detectable by all procedures.

(e) *Relative Redundancy*
The relative redundancy measured by the redundancy numbers r_i (cf. 5) influences the detectability of errors. The theoretical results from robust statistics, especially with reference to ML-type estimation, are only valid for relative redundancies above 0.8, i.e. when the number of observations is larger than five times the number of unknown parameters.

(f) *Number of Unknowns*
The number of unknowns directly influences the algorithmic complexity.

The four procedures can now easily be characterized.

1. *Complete Search*
Complete search checks all, i.e. up to 2^n, possible configurations of good and bad observations to find the optimum solution. The optimization function obviously should contain a cost-term for bad observations in order not to select a minimum of μ observations yielding residuals $e_i = 0$, or the best set of $\mu + 1$ observations allowing one to estimate $\hat{\sigma}_0^2$ with only one redundant observation. Such a penalty may be derived using the principle of minimum description length, thus relying on the negative logarithm mixed distribution (cf. Fig. 3.6).
Obviously complete search is only feasible for a small number n of observations, a small redundancy r or in case the maximum number $b_{\max}$ of

expected errors is small, as the number of possibilities is

$$\sum_{k=0}^{\min(r-1,b_{\max})} \binom{n}{k} < 2^n. \tag{3.34}$$

Implementation requires either approximate values or the invertibility of the model using an iterative or a direct solution technique.

2. *Random Sample Consensus (RANSAC)*
Random sample consensus relies on the fact that the likelihood of hitting a good configuration by randomly choosing a set of observations is large. This probability of finding at least one good set of observations in t trials is $1 - (1 - (1 - \varepsilon)^u)^t$ where u is the number of trials and ε the percentage of errors. For example, for $u = 3$ (spatial resection, fitting circle in the plane) and an outlier rate of 50% at least $t = 23$ trials are necessary, if this probability is to be larger than 95%.
Again, the technique requires approximate values or the invertibility of the model and is only suited for small u.

3. *Clustering*
Clustering consists of determining the probability density function $f_x(y)$ under the assumption that the data represent the complete sample. The mode, i.e. the maximum, of $f_x(x \mid b)$ is used as an estimate. This is approximated by $f_x(x \mid b) \approx \sum_i f_x(x \mid b_s^{(i)})$ where the sum is taken over all or at least a large enough set of subsets b_s of u observations, implicitly assuming these subsets to be independent.
The Hough transformation is a classical example of this technique. Stockman [19] discusses the technique in the context of pose determination, thus for determining the mutual orientation between an image and a model.
Clustering is recommendable for problems with few unknowns, high percentage of gross errors and in cases in which enough data can be expected to support the solution (high relative redundancy).

4. *Maximum-Likelihood-Type Estimation*
Maximum-likelihood-type estimation is based on an iterative scheme. Usually the method of modified weights is used showing the close relation to the classical ML-estimation, where the observations are assumed to be Gaussian distributed. Instead of minimizing $\sum(e_i/\sigma_i)^2$, the sum of a less increasing function $\rho(e_i/\sigma_i)$ is minimized. This can be shown to be equivalent to iteratively weighting down the observations using the weight function $w(x) = \rho'(x)/x$. For convex and symmetric ρ, bounded and monotone decreasing $w(x)(x > 0)$ and a linear model uniqueness of the solution is guaranteed [13]. Since the influence function $\rho'(x)$ [11] stays strictly positive in this case, indicating large errors still influencing the result, nonconvex functions ρ are used.
Most orientation problems are nonlinear and the influence of large errors should be eliminated, thus approximate values are required when using

this ML-type estimation. Further requirements are: moderate sized errors, small percentage of errors and homogeneous design, i.e. large enough local redundancy (no leverage points). The advantage of this technique is its favorable computational complexity being $O(u^3 + nu^2)$ in the worst case allowing it to be used also for large u where sparse techniques may be applied to further reduce complexity.

Without discussing the individual techniques for robust estimation in detail, which would uncover a number of variations and modifications necessary for implementation, the techniques are obviously only applicable under certain – more or less precisely known – conditions. Moreover, specific properties both of the techniques and of the problem to be solved suggest the development of heuristic rules for the application of the various techniques leading to a generic strategy for using robust techniques, which will be discussed in the final section.

3.4   Generic Estimation Procedures

Generic estimation procedures need to choose the technique optimal for the problem concerned and be able to evaluate their performance as far as possible. This section discusses a first step in formalizing strategic knowledge and the mutual role of robust estimates and diagnostic tools.

3.4.1   Rules for Choosing Robust Estimation Techniques

The qualitative knowledge about the four robust estimation techniques discussed in the previous section is collected in Table 3.3. It shows the degree of recommendation for each technique dependent on the eight criteria. These criteria refer to:

- necessary prerequisite (approximate values, direct solution);
- likelihood of success (number of observation, reliability, size and percentage of errors); and,
- computational complexity (number of parameters, speed requirements).

We distinguish four degrees of recommendation:

- "very good". In case all indicated criteria are fulfilled ("and"); the technique can exploit its power and usually is best.
- "good". In case none of the criteria for "bad" is fulfilled; the technique works "not bad".
- "bad". In case one of the indicated criteria is fulfilled ("or"); the technique shows unfavorable properties, so is unreliable or too costly.
- "impossible". In case all indicated criteria are fulfilled ("and"); the technique cannot be used.

Table 3.3. Properties of four techniques for robust estimation.

	Complete search			RANSAC			Clustering			ML-type estimation		
	vg	b	i	vg	b	i	vg	b	i	vg	b	i
Approximate values		–				–	+		–	+		–
Direct solution		–	–	+	–	–	+	–	–			
Many observations	–	+		+			+	–		+		
Few parameters	+	–		+	–		+	–				
High reliability				+						+	–	
Large errors	+			+			+			–	+	
High error rate	+						+			–	+	
Speed unimportant	+	–					+	–		–		

vg = very good (and)
b = bad (or)
i = impossible (and)
    (possible = not(impossible))
+ = feature required
– = feature not required

This knowledge can easily be put into rules, e.g. using PROLOG, together with a few additional rules for qualitative reasoning, e.g. `very-recommendable(X): - good(X), possible(X)` or `impossible(X): - not (impossible(X))`. This allows for the automatic selection of the robust estimation procedure which fits best to the problem at hand.

Example. The determination of the extrinsic parameter of a camera orientation using sets of straight line segments, already mentioned above (example on sensitivity analysis), is performed in several steps.

Step 1 Estimation of the approximate position (two parameters) of the projected model of each set in the image in order to obtain a preliminary set of candidate matches between image and model segments. A sample dialog is given below:

```
Please characterize your problem:
Answer with y=yes, n=no, -=do not know, ?=help

Does a direct solution exist       (Default=n)   ?   y
Are approximate values available   (Default=n)   ?   y
Do you have many observations      (Default=n)   ?   y
Are there many unknown parameters  (Default=n)   ?   n
Is the percentage of errors large  (Default=y)   ?   y
Do you expect large blunders       (Default=y)   ?   y
Is computational speed essential   (Default=y)   ?   n
```

88   Wolfgang Förstner

```
Strongly recommended:
    clustering
    ransac
Not recommended:
    ml type estimation
    complete search
```

Clustering and RANSAC are strongly recommended. ML-type estimation is not recommended as large errors are to be expected. Complete search is not recommended as the number of observations is large.

Step 2 Estimation of good approximate values for the six orientation parameters based on the reference points for each object. Thus only few (point) observations are available. The decision is shown in Table 3.4. As computational speed is made essential in this step and the percentage of errors is large (e.g. three out of eight points) only RANSAC is recommendable.

Step 3 The final cleaning of the observations again refers to the line segments. The system does not have access to a direct solution (e.g. by Horaud et al. [12]) and is required to be fast. Therefore ML-type estimation is highly recommendable.

Obviously the qualitative reasoning may be made more precise:

- The number of observations (few, many) is actually known in a special situation. It influences the density of the cells in clustering, the relative redundancy, specifically the homogeneity of the design and the likelihood of finding a good set in RANSAC.

Table 3.4. The decision of a PROLOG program for the selection of the appropriate robust estimation technique.

	Step 1[1)	Step 2	Step 3
Direct solution	yes	yes	no
Approximate values	yes	yes	yes
Many observations	yes	no	yes
Many unknowns	no	no	no
Many errors	yes	yes	no
Large errors	yes	no	no
Speed essential	no	yes	yes
Very recommendable	clustering RANSAC	–	ML-type
Recommendable	clustering RANSAC	RANSAC	ML-type RANSAC
Not recommendable	ML-type complete search	all except RANSAC	clustering complete search

[1) cf. sample dialog

- The number of unknowns (few, many) also is known in a specific situation and can be used to predict the computational effort quite precisely.
- The homogeneity of the design can usually be approximated in case the number of observations is much higher than the number of the unknowns, which was actually implemented as a rule in the above mentioned PRO-LOG program.
- The size and the percentage of the errors to be expected can be predicted from previous data sets and information which vision algorithms should report for learning their performance.
- The required speed can usually be derived from the specification of the application and be quite rigorously related to the available resources. An example for such a performance prediction in the context of recognition tasks is given by Chen and Mulgaonkar [5].

The final goal of a formalization of tool selection would be to leave the choice of the appropriate estimation procedure to the program, which of course requires the standardization of the input/output relation for the procedures.

3.4.2   Integrating Robust Estimation and Diagnosis

For achieving results which are robust with respect to model errors an integration is necessary which exploits the diagnostic tools and the robust procedures. Diagnostic tools do not influence the estimates and robust estimates only work if the design is homogeneous, requiring a rigorous diagnosis.

Therefore, three steps are necessary:

1. The mensuration design has to be planned in order to guarantee that model errors are detectable and undetectable model errors have only acceptable influence on the result. Diagnostic tools are available for all type of model errors; gross errors, systematic errors or errors in distribution. Strategies for planning, however, are poorly formalized and up to now require at least some interactive effort.
2. Robust estimation techniques may then be used to find an optimal or at least a good estimate for the unknown parameters. This actually can be viewed as an hypothesis generation about the quality (good, bad) of the observations, which is obvious in matching problems which use robust techniques. Any technique may be used which leads to good hypothesis.
3. In the final step the parameters are optimally estimated, e.g. using ML-type estimation based on the decision made in the previous step. The result can then be rigorously checked if it meets the requirements set up in the planning phase. Thus here again the diagnostic tools are used. The result of this analysis provides an objective self-diagnosis which may then be reported to the system in which the estimation procedure is embedded.

The overall quality of the estimation procedure is its ability to correctly predict its own performance, which of course can only be checked empirically [17].

Example. Table 3.5 summarizes the result of 48 image orientations.

The total number of correct and false decisions of the self-diagnosis is split into the cases where the images contained six or more points, i.e. sets of straight line segments, and cases with five or less points. An orientation was reported as correct if the empirical and the theoretical sensitivity factors $\Delta_i f/\sigma_f$ and $\Delta_{0i} f/\sigma_f$ (cf. (3.28) and (3.31)) and the standard deviations of the result were acceptable ($\lambda_{max} < 1$, cf. (3.12)).

Table 3.5. The result of an extensive test of orienting 48 aerial images, # of cases, in brackets: for ≥ 6 points/image and for ≤ 5 points/image.

		Report of self-diagnosis	
		correct	wrong
Reality	correct	46(39/7) (correct decision)	0(0/0) (false positives)
	wrong	1(0/1) (false positives)	1(0/1) (correct decisions)

46 out of 48 orientations were correct and this was reported by the self-diagnosis. In one case the orientation was incorrect, which was detected by the analysis. This appeared in an orientation with only four points, thus only one redundant point. Therefore altogether in 47 out of 48, i.e. in 98% of all orientations the system made a correct decision. In about half of the cases (22 out of 46) the RANSAC procedure was able to identify errors which occurred during the clustering and correct the result of the clustering, which was repeated with this a priori knowledge.

One orientation failed without being noticed by the system, which corresponds to 2% false positives. This was an orientation with only five points.

The orientation of the 48 images was based on 362 clusterings of model and image line segments. 309, thus 85%, were correct. As the errors in clustering are either completely wrong and therefore eliminated from the further processing or are wrong by a small amount, it is quite likely that two clusterings are incorrect by only a small amount, which may not be detectable by the RANSAC or the robust ML-type estimation, mimicking a good orientation. Therefore the existence of one false positive is fully acceptable.

The result achieved in this test is a clear reason to require at least six points, i.e. sets of straight edges, for a reliable orientation in this application. As can be seen in the table, all 39 orientations not only could correctly be handled by the automatic system, but actually lead to correct orientation parameters.

This example reveals the diagnostic tools to be extremely valuable for a final evaluation of an automatic procedure containing robust estimation procedures as parts.

3.5   Conclusions

The goal of this chapter was to collect the tools from robust statistics and diagnostics necessary for building fully automatic image analysis procedures, specifically orientation procedures. The theory available seems to be sufficient for achieving a high degree of self-diagnosis and for implementing generic strategies based on knowledge about the specific properties of the different estimation techniques.

In all cases the general strategy for achieving results of high quality consists of four steps:

1. Planning the mensuration configuration using the diagnostic tools for precision and sensitivity garanteeing robust estimation techniques in step 3 to be applicable and the result to be evaluated internally.
2. Mensuration according to the planned configuration which itself gives a clear indication of its quality.
3. Robust estimation with any of the available techniques leading to a hypothesis of a good result.
4. Final evaluation based on the result of an optimal estimation of the parameters and a check on how far the quality intended in the planning stage actually is reached.

The examples give a clear indication that these tools can be used to advantage even in comparatively complex situations.

There are still some questions open:

- The analysis of the precision of the result is based on the comparison of the actual with the required covariance matrix. Generating meaningful criterion matrices requires proper *modeling of the user needs* making the specification of the accuracy requirements a nontrivial problem.
- The effect of systematic errors (biases in the model) can be analysed with the same techniques. However, the *search space for identifying undetectable systematic errors is large*, due to the unknown interference between the different causes for such errors.
- The planning of experiments may be based on the techniques collected in this chapter. Automating the planning, as it may occur in active vision, however, requires the development of strong *strategies* for finding optimal or at least satisfying observation *configurations*.

- The self-diagnosis, based on the precision and sensitivity analysis, does not give indications on the *probability* of the result to be correct. This would enable the calling routine to react in a more specific manner or to use this probability for further inference.

It would, however, be of great value if all orientation procedures would offer at least the available measures for making a proper self-diagnosis in order to objectify the quality of the very first steps within image analysis.

Appendix A
Algebraic Expression for the Normal Equations
of Spatial Resection with Four Parts
in Symmetric Position.

Let points P_i with coordinates $(x_i, y_i, z_i), i = 1, \ldots, n$ in the camera system be given and observed in the image. The linearized observation equations for the image coordinates (x', y') depending on the six orientation parameters, namely the rotation angles ω, φ, κ and the position (x_0, y_0, z_0) of the projection center can be expressed as

$$dx_i' = -\frac{c}{H}z_0 dx_0 - \frac{x_i'}{H}z_0 dx_0 - \frac{x_i' y_i'}{c}d\omega + c(1 + \frac{x_i'^2}{c^2})d\phi + y_i' d\kappa \quad (3.35)$$

$$dy_i' = -\frac{c}{H}z_0 dy_0 - \frac{y_i'}{H}z_0 dy_0 - c(1 + \frac{y_i'^2}{c^2})d\omega + \frac{x_i' y_i'}{c}d\varphi - x_i' d\kappa \quad (3.36)$$

valid for each image point $P'(x_i', y_i')$. c is the camera constant.

In case $n = 4$ image points lie in symmetric position $(\pm d, \pm d)$ in the image (cf. Fig. 3.7) and the z-coordinates of the points $P_i(x_i, y_i, z_i)$ in the coordinate system of the camera are equal to $H = z_i$, we can collect the coefficients of the 8×6 matrix A as in Table 3.6.

The algebraic expression for the normal equation matrix $N = A^T P A$ assuming weights 1 for the observations is given by

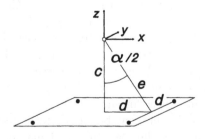

Fig. 3.7. The normalized situation for spatial resection used as a reference for precision.

Table 3.6.

i		dx_0	dy_0	dz_0	$d\omega$	$d\phi$	$d\kappa$
1	$x' = +d$	$-\frac{c}{H}$	0	$-\frac{d}{H}$	$-\frac{d^2}{c}$	$\frac{e^2}{c}$	$+d$
	$y' = +d$	0	$-\frac{c}{H}$	$-\frac{d}{H}$	$-\frac{e^2}{c}$	$\frac{d^2}{c}$	$-d$
2	$x' = -d$	$-\frac{c}{H}$	0	$+\frac{d}{H}$	$+\frac{d^2}{c}$	$\frac{e^2}{c}$	$-d$
	$y' = +d$	0	$-\frac{c}{H}$	$-\frac{d}{H}$	$-\frac{e^2}{c}$	$\frac{d^2}{c}$	$-d$
3	$x' = -d$	$-\frac{c}{H}$	0	$+\frac{d}{H}$	$-\frac{d^2}{c}$	$\frac{e^2}{c}$	$-d$
	$y' = -d$	0	$-\frac{c}{H}$	$+\frac{d}{H}$	$-\frac{e^2}{c}$	$\frac{d^2}{c}$	$+d$
4	$x' = +d$	$-\frac{c}{H}$	0	$-\frac{d}{H}$	$+\frac{d^2}{c}$	$\frac{e^2}{c}$	$+d$
	$y' = -d$	0	$-\frac{c}{H}$	$+\frac{d}{H}$	$-\frac{e^2}{c}$	$-\frac{d^2}{c}$	$+d$

$$N = \begin{pmatrix} 4\frac{c^2}{H^2} & 0 & 0 & 0 & -4\frac{e^2}{H} & 0 \\ 0 & 4\frac{c^2}{H^2} & 0 & 4\frac{e^2}{H} & 0 & 0 \\ 0 & 0 & 6\frac{d^2}{H^2} & 0 & 0 & 0 \\ 0 & 4\frac{e^2}{H} & 0 & 4\frac{e^4}{c^2}(1+\frac{d^4}{e^4}) & 0 & 0 \\ -4\frac{e^2}{H} & 0 & 0 & 0 & 4\frac{e^4}{c^2}(1+\frac{d^4}{e^4}) & 0 \\ 0 & 0 & 0 & 0 & 0 & 8d^2 \end{pmatrix}. \tag{3.37}$$

Discussion

1. The normal equation matrix is sparse. It collapses to two diagonal elements and two 2×2 matrices. This allows algebraic inversion (which may be used for a direct solution of the orientation in real time applications).
2. The correlation between x_0 and $d\phi$ (y-rotation), y_0 and $d\omega$ (x-rotation) is given by

$$\rho_{y_0\omega} = -\rho_{x_0\phi} = \frac{N_{24}}{\sqrt{N_{22} \cdot N_{44}}} = \frac{4\frac{e^2}{H}}{\sqrt{4\frac{c^2}{H^2} \cdot 4\frac{e^4}{c^2}(1+\frac{d^4}{e^4})}} = \frac{1}{\sqrt{1 + \sin^4\frac{\alpha}{2}}} \tag{3.38}$$

as $d/e = \sin\alpha/2$ (cf. Fig. 3.7).
3. Taking the square root $N^{1/2}$ of N is trivial for the diagonal elements for dz_0 and $d\kappa$ and requires us to take the square root $T^{1/2}$ of two 2×2 matrices T which can easily be determined using the eigenvalue decomposition $T = D\Lambda D^{\mathrm{T}}$ yielding

$$T^p = D\Lambda^p D^{\mathrm{T}} \tag{3.39}$$

for $p = 1/2$ or, as needed in (3.10) for $p = -1/2$ ($D^{\mathrm{T}} = D^{-1}$ and $\Lambda = \mathrm{Diag}(\lambda_1, \lambda_2)$).

94   Wolfgang Förstner

References

1. W. Baarda. S-Transformations and Criterion Matrices. Netherlands Geodetic Commission New Series 5(1), Delft, 1973
2. W. Baarda. Statistical Concepts in Geodesy. Netherlands Geodetic Commissison, New Series, 2(4), Delft, 1967
3. W. Baarda. A Testing Procedure for Usen in Geodetic Networks. Netherlands Geodetic Commissison, New Series, 5(1), Delft 1968
4. R.C. Bolles, M.A. Fischler. Random Sample Consensus: A Paradigm for Model Fitting with Applications to Image Analysis and Automated Cartography. Comm. of ACM 24(6), 381–395, 1981
5. C.-H. Chen, P. G. Mulgaonkar. Robust Vision-Programs Based on Statistical Feature Measures. Proc. International Workshop on Robust Computer Vision, pp. 39–56, Seattle, October 1990
6. R.D. Cook, S. Weisberg. *Residuals and Influence in Regression.* Chapmann & Hall, London, 1982
7. R. Deriche, O. Faugeras. Tracking Line Segments. Lecture Notes in Computer Science 427, 259–268, Springer, 1990
8. W. Förstner. Reliability Analysis and Parameter Estimation in Linear Models with Application to Mensuration Problems in Computer Vision. CVGIP 40, 87–104, 1987
9. W. Förstner. Reliability and Discernability of Extended Gauss-Markov Models. Deutsche Geod. Komm. A98, 79–103, München 1983
10. W. Förstner. *Uncertain Geometric Relationships and their Use for Object Location in Digital Images.* Institut für Photogrammetrie, Universität Bonn, Tutorial, 1992
11. F.R. Hampel, E.M. Ronchetty, P.J. Rousseeuw, W.A. Stahel. *Robust Statistics: The Approach Based on Influence Functions.* Wiley, 1986
12. R. Horaud, B. Conio, O. Leboulleux, B. Lacolle. An Analytic Solution for the Perspective 4-Point Problem. Proc. CVPR 1989, pp. 500–507, San Diego, CA, 1989
13. P.J. Huber. Between Robustness and Diagnostics. In: *Directions in Robust Statistics and Diagnostics,* Part I, W. Stahel and S. Weisberg (eds.), Springer Verlag, pp. 121–130, 1991
14. P. J. Huber. *Robust Statistics.* Wiley, New York, 1981
15. G. Roth, M.D. Levine. *Random Sampling for Primitive Extraction.* Proc. International Workshop on Robust Computer Vision, Seattle, pp. 352–366, October 1990
16. P.J. Rousseeuw, A.M. Leroy. *Robust Regression and Outlier Detection.* Wiley, 1987
17. W. Schickler. Feature Matching for Outer Orientation of Single Images Using 3-D Wireframe Controlpoints. International Archive of Photogrammetry & Remote Sensing 29, ISPRS Comm. III, Washington D. C., 1992
18. M. Sester, W. Förstner. Object Location Based on Uncertain Models. Informatik Fachberichte 219, 457–464, 1989
19. G. Stockman. Object Recognition and Localization via Pose Clustering. CVGIP 40, 361–387, 1987
20. Wrobel 1995

4 Photogrammetric Camera Component Calibration: A Review of Analytical Techniques

Clive S. Fraser

Summary

As the field of computer vision advances, more and more applications are calling for enhanced metric performance. Central to any consideration of metric accuracy is the subject of camera system calibration. Analytical camera calibration techniques developed over the past four decades in photogrammetry are very applicable to the video cameras used in today's machine vision systems. This chapter reviews modern analytical camera calibration techniques employed by photogrammetrists and discusses the potential of these methods for video cameras (principally CCD cameras). Analytical restitution is briefly overviewed, the parameterization of the calibration components is reviewed, and the methods of parameter determination are discussed. The use of photogrammetric calibration techniques offers a means of significantly improving the accuracy of spatial position and orientation information in computer and machine vision systems.

4.1 Introduction

In a modern photogrammetric system for industrial measurement, image mensuration accuracies can exceed 1:200 000 of the camera's field of view, and the resulting object space positional accuracies can surpass 1:1 000 000 of the principal dimension of the object [1]. To achieve such measurement resolution it is critical to have a comprehensive understanding of the geometric relationship between the image space of the data acquisition sensor(s) and the object space. Central to this understanding is the aspect of system calibration. Through a combination of analytical calibration techniques and specialized camera technology, photogrammetrists have continued to advance the geometric fidelity of the photogrammetric restitution process. System calibration is thus a very necessary, though not sufficient condition for accurate 3D measurement by means of photogrammetric triangulation.

While it is certainly a fact that photogrammetrists employ highly accurate, fully metric cameras to perform the more exacting measurement tasks, use of nonspecialized, even "amateur" camera equipment is not precluded for

Springer Series in Information Sciences, Vol. 34
Calibration and Orientation of Cameras in Computer Vision
Eds.: Gruen, Huang © Springer-Verlag Berlin Heidelberg 2001

applications demanding low to moderate accuracies of, say, 1:1000 to 1:20 000. Analytical calibration techniques have gone a long way towards providing the means to compensate for geometric perturbations of the imaging process. Indeed, with two notable exceptions, namely the degree of focal plane unflatness and interior orientation instability, the same set of parameters affecting measurement accuracy is applicable to both metric and nonmetric camera systems.

Within the machine vision community, with its interests in video imaging, the subject of system calibration has not thus far received the same attention as in photogrammetry. For the application of any vision system for metric measurement purposes, however, there is still the need for a calibration model to form part of the quantitative image analysis. In the context of calibration, and especially geometric calibration, video cameras display both similarities and differences to those which are film-based. A CCD matrix array camera is arguably metric if it exhibits a stable interior orientation (stable "intrinsic" parameters) and has a digital output (to circumvent pixel positioning errors associated with standardized video signals). Recently reported photogrammetric evaluations of digital CCD cameras such as the Videk Megaplus and the 1K × 1K camera from Thomson have produced object space triangulation accuracies in the range of 1:50 000–1:80 000, which certainly constitutes high-precision metric performance [2,3]. Accuracies approaching this level have also been attained with "standard" 512 × 512 CCD cameras employing external pixel clock synchronization [4,5].

Vidicon cameras on the other hand fit firmly into the nonmetric category due to instabilities (e.g. drift) in the imaging process. Nevertheless, photogrammetric calibration techniques can assist in improving the metric performance of such cameras in instances where, for example, image plane errors due to lens distortion are much larger than the inherent image measurement uncertainties and instabilities. In modern-day, close-range photogrammetry, analytical camera calibration techniques are the norm. Laboratory methods involving goniometers and optical bench techniques are very rarely applied to film cameras. Laboratory methods have, however, found limited application for the calibration of video cameras. Burner et al. [6] report on a technique which utilizes a laser-illuminated displaced reticle and collimator for the determination of parameters of interior orientation. Distortion characteristics are also recovered via a combination of laboratory and analytical methods.

One argument in favor of laboratory approaches is that they give a better insight into the behavior of individual calibration parameters. There are, however, disadvantages which include the need for specialized equipment and facilities, and the fact that the methods employ time-consuming alignment and measurement. Moreover, there is the question of applicability of laboratory calibration to the physical environment in which the sensor system is to be employed. Perhaps the principal reason why laboratory calibration is out of favor in modern close-range photogrammetry is that analytical calibration techniques such as test-range and self-calibration are such comprehensive and easy to use on-the-job techniques which are applicable for both film and video camera systems.

Analytical calibration techniques for video camera and stereo imaging systems are not presently the exclusive domain of photogrammetrists. As the demands increase for improved metric performance in 3D vision systems, so researchers within the vision community have turned their attention to this problem. In Tsai [7] a number of categories of existing camera calibration techniques are reviewed. In the paper, an ideal calibration technique is defined as one which is "autonomous, accurate, efficient, versatile and requires only common off-the-shelf cameras and lenses". The author then implies that photogrammetric calibration techniques are less than optimal in the pursuit of these goals. An aim of this chapter is to highlight the very attributes of the photogrammetric calibration process which are indeed ideal in terms of the requirements above.

In this presentation analytical photogrammetric camera calibration will be reviewed. While the techniques to be discussed have been developed primarily with film (and plate) cameras in mind, they are also applicable in the vision environment of fast, real-time data processing, and especially in the pre- and post-calibration of the video cameras. Camera component calibration through photogrammetric techniques offers the potential of significantly improving metric performance without adversely affecting computational load. The calibration techniques reviewed are presented in the framework of analytical photogrammetric restitution and it is on this subject that the discussion will be commenced. The review will be kept general in nature, with no reference being made to specific vision calibration categories such as active vision sensors in eye-in-hand robotics configurations. The photogrammetric nature of the discussion, however, will mean that optical triangulation systems are emphasized.

Before addressing analytical restitution, an important common thread that binds all photogrammetric calibration approaches is worth noting. Put simply, this could be said to be the comprehensive geometric (and radiometric where appropriate) modelling of the image space of the camera system, independent of the target scene. The resulting calibration is then without regard to the object space and is largely independent of the geometric configuration of the camera network employed for the particular 3D measurement task at hand.

4.2   Analytical Restitution

4.2.1   Interior and Exterior Orientation

The process of analytical photogrammetric restitution involves the perspective transformation between image and object space. This optical triangulation process is illustrated in Fig. 4.1. The fundamental problem is to determine XYZ object space coordinates given the corresponding xy image coordinates on two or more photographs (or digital images). To solve the problem, the spatial direction of each of the intersecting rays and the location of the perspective center for each exposure must be established with respect to the XYZ coordinate system. The six parameters involved, three orientation angles and three camera station coordinates, describe what is termed the exterior orientation of the image. (The term "extrinsic parameters" has been used on occasion for exterior orientation in the machine vision literature.)

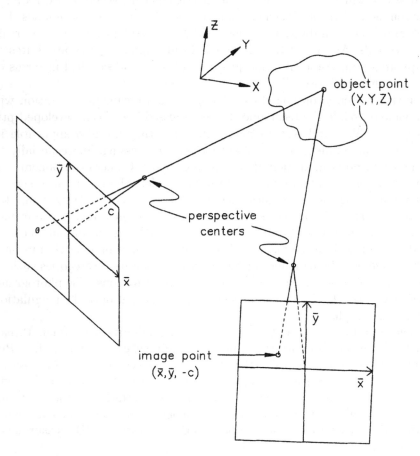

Fig. 4.1. Photogrammetric triangulation.

The relationship between the perspective center and the image coordinate system is also required and this is defined by the camera's interior orientation. There are three parameters of interior orientation, namely the camera principal distance, c, and the coordinates (x_0, y_0) of the principal point. Through reference to Fig. 4.2, it can be seen that the principal distance is the perpendicular distance from the perspective center to the focal plane, and the principal point is the point at which the optical axis intersects the image plane. While the origin of the xy photo coordinate system should ideally be coincident with the principal point, this is rarely the case. Thus, the principal point offsets x_0 and y_0 define the shift between the origin of the xy system, defined by fiducial (image reference) marks, and the principal point. These two parameters of interior orientation also apply to a CCD matrix array camera where, instead of a fiducial system, a central row and column are employed to define the origin of the xy image coordinate system. Values of up to 0.5 mm for x_0 and y_0 are not uncommon in CCD cameras.

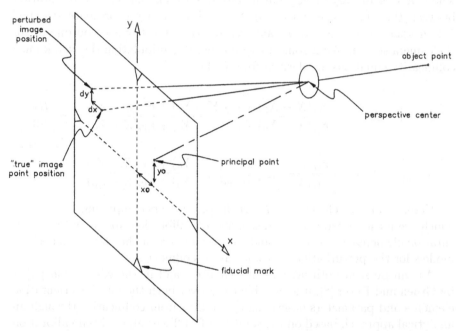

Fig. 4.2. Interior orientation and the influence of perturbations to collinearity.

4.2.2   Collinearity Equations

When taking a photograph with a "perfect" camera system the object point, perspective center and image point will all lie along the same straight line. Perturbations to the imaging process, however, give rise to departures from collinearity which manifest themselves as departures Δx and Δy of the image point from its true position on the focal plane, as shown in Fig. 4.2. These perturbations are both real and "mathematical" in nature. In the real category there is lens distortion and atmospheric refraction, whereas in the mathematical group the adoption of incorrect values for interior orientation elements would be included. In simplest terms, the calibration of a photogrammetric imaging system involves the modeling of these perturbations to facilitate corrections for the induced image coordinate shifts Δx and Δy. That is, the actual image point is shifted so as to satisfy the collinearity requirement.

The collinearity condition is implicit in the perspective transformation between image and object space, which is of the form:

$$\begin{pmatrix} x - x_0 + \Delta x \\ y - y_0 + \Delta y \\ -c \end{pmatrix} = \lambda R \begin{pmatrix} X - X_0 \\ Y - Y_0 \\ Z - Z_0 \end{pmatrix}. \tag{4.1}$$

where R is a unitary-orthogonal matrix describing the relative orientation between the xyz image space and XYZ object space coordinate systems, λ is a scale factor and X_0, Y_0 and Z_0 are the object space coordinates of the camera station. After some straightforward manipulation the well-known collinearity equations are derived from (4.1):

$$x - x_0 + \Delta x = -c\frac{r_{11}(X - X_0) + r_{12}(Y - Y_0) + r_{13}(Z - Z_0)}{r_{31}(X - X_0) + r_{32}(Y - Y_0) + r_{33}(Z - Z_0)} = -c\frac{R1}{R3}$$

$$\tag{4.2}$$

$$y - y_0 + \Delta y = -c\frac{r_{21}(X - X_0) + r_{22}(Y - Y_0) + r_{23}(Z - Z_0)}{r_{31}(X - X_0) + r_{32}(Y - Y_0) + r_{33}(Z - Z_0)} = -c\frac{R2}{R3}$$

Here, r_{ij} are the elements of R. Of the parameters comprising (4.2), those which are most relevant to the discussion of calibration are the interior orientation elements x_0, y_0 and c, and the parameters of the as yet unspecified models for the perturbations Δx and Δy to collinearity.

As will be seen both later in this chapter, and in the companion paper by Gruen and Beyer [8], it is possible to recover both the interior orientation elements and parameters describing departures from collinearity through an analytical approach based on the solution of (4.2). Analytical self-calibration and test-range calibration embody this approach. Self-calibration implies the simultaneous solution of all parameters forming the collinearity equations: exterior and interior orientation, calibration coefficients and object space target point coordinates. This generalized relative orientation of all bundles of rays can be performed without any a priori knowledge of 3D coordinate or scale information regarding the object space target field.

Test-range calibration, as the name implies, requires the provision of a suitable control field comprising pre-surveyed targets with known XYZ coordinates. Restitution is then a two-step process. The exterior and interior orientation elements, and camera calibration parameters are first recovered in what amounts to a self-calibrating resection. Any subsequent triangulation of object points would then be through intersection based on the computed exterior orientation information, though the intersection phase is rarely carried out in conjunction with this "on-the-job" calibration operation. Test ranges per se are also employed in analytical self-calibration to aid in the verification of derived camera parameters. In this discussion use of the term test-range will imply the necessity of an object space control field in the calibration procedure; self-calibration has no such requirement.

The first necessity for the analytical calibration process is a parameterization of the perturbation functions Δx and Δy. Before proceeding to this step, however, it is useful to note that the collinearity equations, (4.2), do not offer the only functional model for analytical restitution. There are alternatives employed in close-range photogrammetry, the most prominent of which will now be briefly reviewed.

4.2.3   The DLT

The direct linear transformation (DLT) facilitates a perspective transformation between two-dimensional image space data and three-dimensional object space. The DLT, as originally formulated by Abdel Aziz and Karara [9], combined into a single linear model the 2D affine transformation from film reader to image coordinates and the transformation from image to 3D object space coordinates via the collinearity model detailed above. The basic projective equations of the DLT are as follows:

$$x + \Delta x = \frac{L_1 X + L_2 Y + L_3 Z + L_4}{L_9 X + L_{10} Y + L_{11} Z + 1}$$

$$y + \Delta y = \frac{L_5 X + L_6 Y + L_7 Z + L_8}{L_9 X + L_{10} Y + L_{11} Z + 1}$$

(4.3)

where x and y may be either comparator or image (with respect to the fiducial center) coordinates, or pixel coordinates not necessarily referenced to the principal point. The 11 parameters $L_1 - L_{11}$ can be physically interpreted in terms of the interior and exterior orientation of the image, though the parameters are not strictly equivalent to the perspective parameters of the collinearity equations. Moreover, linear dependencies exist between the 11 parameters, a factor which is taken into account in an alternative DLT formulation of Bopp and Krauss [10] in which two orthogonality constraints are imposed in the 11 parameter transformation.

Application of the DLT has proved popular for the restitution of non-metric photography since no a priori knowledge of the interior orientation elements (x_0, y_0, c) is required. In the digital camera context, the DLT can offer two advantages for space resection not exhibited by the collinearity approach. Firstly, a noniterative, direct solution which is independent of initial parameter estimates is achieved, and this offers the potential of fast computation. Secondly, the affine/shear image coordinate correction implicit in the model is quite appropriate for CCD sensors [24]. On the other hand, the

DLT as a resection model requires more object space control points than the collinearity approach, and the control should be well distributed in three dimensions (a cube configuration, for example).

The computational models of collinearity and the DLT display different characteristics in terms of numerical stability, parameter covariances, least-squares solution convergence, general computational effort and object point control. When considering camera self-calibration or test-range calibration, careful attention must be paid to distinctions between the two methods. Analytical calibration methods are independent of the photogrammetric restitution model to some degree, however. In this treatment, a more generic approach will be adopted to alleviate the need to dwell too deeply on the distinctive characteristics of the different restitution approaches. The discussion of analytical restitution will be confined to the collinearity model since this is both the most flexible and most widely adopted in close-range photogrammetry.

4.3   Parameterization of Departures from Collinearity

4.3.1   Sources of Perturbation

In seeking appropriate parameters for the functions Δx and Δy it is first necessary to consider the four principal sources of departures from collinearity which are physical in nature. These are symmetric radial distortion, decentering distortion, focal plane unflatness and in-plane image distortion. The net image displacement at a point will amount to the cumulative influence of each of these perturbations. Thus,

$$\Delta x = \Delta x_r + \Delta x_d + \Delta x_u + \Delta x_f$$

$$\tag{4.4}$$

$$\Delta y = \Delta y_r + \Delta y_d + \Delta y_u + \Delta y_f$$

where the subscript r is for radial distortion, d for decentering effects, u for out-of-plane unflatness influences and f for in-plane image distortion. The relative magnitude of each of the four image coordinate perturbations depends very much on the nature of the camera system being employed.

In the days of analog photogrammetry metric close-range cameras kept the magnitudes of Δx and Δy small by using specially designed, very low distortion lenses, and glass plates which eliminated distortion but not necessarily unflatness effects. Modern metric cameras for analytical industrial photogrammetry are designed to keep out-of-plane and in-plane film distortion influences to a minimum through the use of vacuum platens and reseaux. Unflatness effects would arise with CCD cameras through chip bowing or the "crinkling" of thin wafers, whereas in-plane image distortion can be introduced through electronic influences such as line jitter.

4.3.2   Radial Distortion

Symmetric radial distortion in analytical photogrammetry is universally represented as an odd-ordered polynomial series, as a consequence of the nature of Seidel aberrations:

$$\Delta r = K_1 r^3 + K_2 r^5 + K_3 r^7 + \ldots \qquad (4.5)$$

where K_i are termed the coefficients of radial distortion and r is the radial distance from the principal point, i.e.

$$r^2 = \bar{x}^2 + \bar{y}^2 = (x - x_0)^2 + (y - y_0)^2 \qquad (4.6)$$

For the majority of medium-angle, non-photogrammetric lenses employed in today's metric and nonmetric close-range cameras, the third-order term is sufficient to account for the induced aberrations. For wide-angle lenses, higher order terms (very rarely above seventh order) are often required to adequately model lens distortion. For low to medium accuracy CCD camera applications, use of the K_1 term alone will usually suffice since the most commonly encountered distortion profile is that of third-order barrel distortion. Inclusion of the K_2 and K_3 terms might be warranted for higher accuracy applications. In either case the decision as to the order of the correction function can be made on the basis of statistical tests of significance of the individual and often highly correlated radial distortion coefficients.

The distortion profile Δr associated with a particular principal distance value, c, is termed the Gaussian distortion profile. The radial distortion curve may also be "balanced" so as to pass through zero at some chosen radial distance. This is achieved through the introduction of a linear term to (4.5) and an associated change, Δc, to the principal distance, c. Sample Gaussian and balanced distortion profiles for a Videk Megaplus CCD camera with 20 mm lens are shown in Fig. 4.3. It is important to note that in applications of the collinearity equations, these two distortion profiles are projectively equivalent. The balancing operation effectively tilts the abscissa of the Gaussian profile, while c is incremented or decremented to compensate for this linear tilt.

The projective coupling between principal distance and lens distortion gives rise to an interesting and fortuitously beneficial feature in the self-calibration of selected CCD cameras. It is not uncommon for CCD cameras to utilize only a modest portion of the available field of view of the lens selected for the camera. Thus, in some circumstances the radial distortion profile for this central lens region may not depart, for all practical purposes, from a linear function of the form $\Delta r = K_0 r$. This linear profile can in turn be completely compensated for by a change in principal distance. The net outcome of the calibration process is then that for the calibrated principal distance the lens shows no significant radial distortion [3].

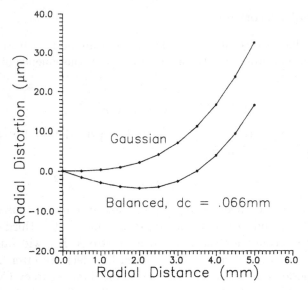

Fig. 4.3. Gaussian and "balanced" radial distortion (Δr) profiles for a Videk Megaplus CCD camera with 20 mm lens.

From (4.5) the necessary radial distortion corrections for the x and y image coordinates follow as

$$\Delta x_r = \frac{\bar{x}}{r}\Delta r$$

$$\Delta y_r = \frac{\bar{y}}{r}\Delta r \;.$$

(4.7)

For a fixed-focus lens one needs only a single set of coefficients K_i. Radial distortion, however, varies both with focussing and within the photographic field for a particular focus [11–13]. The variation with focus is predictable and so long as the distortion coefficients for two focal settings are known the applicable K_i-values for any other principal distance can be computed. The amount of variation changes dramatically for different lens types, from virtually zero to 0.1 mm or more. Figures 4.4–4.5 present two examples. In the first, Fig. 4.4, the variation of distortion with focussed distance is shown for a Schneider 120 mm lens mounted in a GSI CRC-2 medium format photogrammetric film camera. In Fig. 4.5 the distortion variation is illustrated for a Schneider 65 mm lens in the same camera.

With the majority of small format lenses of short focal length utilized with CCD cameras it is reasonable to assume that the magnitude of radial distortion will be significantly greater than the variation of distortion with both focussing and within the photographic field. The implication then for vision systems of low to moderate triangulation accuracy is that distortion variation influences can be ignored. The detrimental effects of neglecting to correct for lens distortion, however, can be significant for cameras exhibiting large radial distortion.

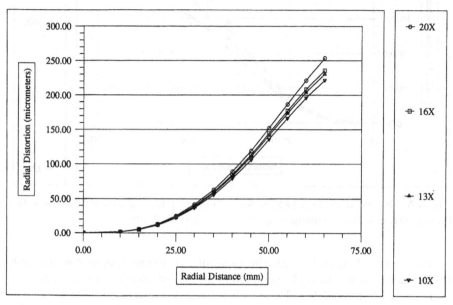

Fig. 4.4. Radial distortion profiles for a Schneider 120 mm lens for focussed distances corresponding to image scales of 10×, 13×, 16× and 20×.

4.3.3   Decentering Distortion

A lack of centering of lens elements along the optical axis gives rise to a second category of lens distortion which has metric consequences in analytical restitution, namely decentering distortion. The misalignment of lens components causes both radial and tangential image displacements which can be modeled by correction equations due to Brown [14]:

$$\Delta x_d = P_1 \left(r^2 + 2\bar{x}^2 \right) + 2P_2 \bar{x}\bar{y}$$

$$\Delta y_d = P_2 \left(r^2 + 2\bar{y}^2 \right) + 2P_1 \bar{x}\bar{y} \ . \tag{4.8}$$

A useful means of representing decentering distortion is via the profile function, $P(r)$, which is given as

$$P(r) = \left(P_1^2 + P_2^2 \right)^{1/2} r^2 \ . \tag{4.9}$$

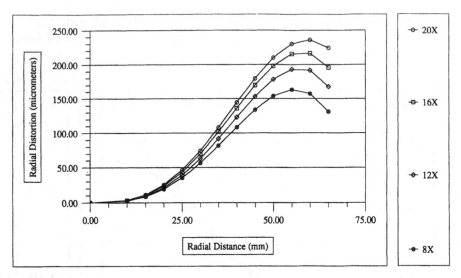

Fig. 4.5. Radial distortion profiles for a Schneider 65 mm lens for focussed distances corresponding to image scales 8×, 12×, 16× and 20×.

For a typical "quality" lens the value of the decentering profile function rarely exceeds a few tens of micrometers at the extremities of the image format, and is often much less. Decentering distortion also varies with focussing, but the resulting image coordinate perturbations are typically small and the distortion variation is universally ignored in analytical photogrammetry.

Although it is not obvious from (4.8), there is a strong projective coupling between the decentering distortion parameters P_1 and P_2 and the principal point offsets x_0 and y_0. This correlation has practical consequences for it means that to a significant extent decentering distortion effects can be compensated for by a shift in the principal point. The projective compensation can usually be anticipated with CCD cameras and hence a self-calibration may indicate that the lens is free of decentering distortion. This feature will be again touched upon in the sections dealing with test-range calibration and self-calibration.

4.3.4   Focal Plane Unflatness

Systematic image coordinate errors due to focal plane or film unflatness constitute a major factor limiting the accuracy of the photogrammetric triangulation process, especially where nonmetric cameras are involved. Unflatness effects are illustrated in Fig. 4.6, where it can be seen that the induced radial image displacement Δr_u is a function of the incidence angle of the imaging ray. Thus, long focal length, narrow-angle lenses are much less influenced by out-of-plane image deformation than short focal length, wide-angle lenses. Unfortunately, many machine vision systems employ wide-angle lenses to

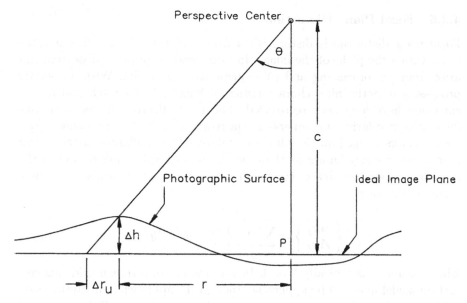

Fig. 4.6. Image displacement due to focal plane unflatness.

achieve a workable field of view in cameras with CCD arrays of small format.

In high-precision industrial photogrammetry, the focal plane topography of a metric film camera can be measured directly [15]. From a series of "spot heights" a polynomial surface model is computed, and the associated image coordinate corrections are then made according to the equations

$$
\left\{ \begin{matrix} \Delta x_u \\ \Delta y_u \end{matrix} \right\} = \left\{ \begin{matrix} \bar{x}/r \\ \bar{y}/r \end{matrix} \right\} \sum_{i=0}^{n} \sum_{j=0}^{i} a_{ij} \bar{x}^{(i-j)} \bar{y}^{(j)} \tag{4.10}
$$

where a_{ij} are the coefficients of the polynomial surface model and the order is typically restricted to $n = 3$ or 4, depending on the number and distribution of "height" points.

The applicability of this approach to CCD matrix arrays is uncertain (at least to this author). Whereas a low-order continuous function may adequately model chip bowing, it is doubtful that crinkling could be successfully modeled in this fashion. Indeed the CCD array may exhibit a degree of planarity that does not warrant any unflatness correction. However, focal plane unflatness should not be ignored; at an incidence angle of 45° a departure from planarity of 10 micrometers will give rise to an image displacement of the same magnitude. Moreover, the influence of unflatness is most insidious in that it invariably leads to significant accuracy degradation in the object space, without aggravating the magnitude of triangulation closures.

4.3.5   Focal Plane Distortion

Focal plane distortion is distinguished from unflatness by the fact it takes place within the plane of the image. In film cameras, in-plane distortion can arise from the processing and subsequent storage of film. With automatic processing of aerial film, characteristics of longitudinal stretch and lateral shrinkage have long been recognized. The most effective means to control distortion in metric film cameras is via reseaux, which are an array of reference marks imaged on the film. The relative xy coordinates of the reseau marks are precisely known at the time the photograph is taken, and so the measured image coordinates can be analytically corrected for distortion by a model of the form

$$\left\{ \begin{matrix} \Delta x_f \\ \Delta y_f \end{matrix} \right\} = \sum_{i=0}^{n} \sum_{j=0}^{i} \left\{ \begin{matrix} b_{ij} \\ c_{ij} \end{matrix} \right\} \bar{x}^{(i-j)} \bar{y}^{(j)} \qquad (4.11)$$

where, again, n is typically 3 or 4. In the absence of a reseau grid this correction model may still be applicable, though the approach of analytical self-calibration then offers the only means to determine the coefficients b_{ij} and c_{ij}.

One encouraging feature of CCD arrays is the high positional integrity of the pixel elements. This would normally indicate that distortion is not a problem with CCD cameras, and indeed it appears not to be for digital CCDs in which A/D conversion occurs in the camera, and image data is output digitally (e.g. the Videk Megaplus and the ProgRes 3000 from Kontron). The distortion which arises in an analog CCD camera is chiefly attributable to A/D conversion (especially to pixel clock nonsynchronization) and video signal transmission. Common symptoms are line jitter and nonlinearities within the scale of the horizontal or x-axis of the image. Recent research [16–18] shows that while these distortions can be large enough to significantly influence measurement accuracies with analog CCD cameras, they can be rendered metrically negligible through pixel synchronous A/D conversion and due attention to aspects such as camera warm-up and power supply fluctuations.

4.3.6   A Practical Model for In-Plane and Out-of-Plane Effects

It should be clear from the terms comprising the correction equations for out-of-plane image deformation (4.10), and in-plane distortion (4.11), that their combination to form a single model for unflatness and distortion effects, through a merging of the equations, will lead to a duplication of terms, with resulting linear dependencies. If the correction coefficients are determined by independent means, this issue is of limited consequence. If, on the other hand, it is desired to recover the coefficients a_{ij}, b_{ij} and c_{ij} via a direct solution to the collinearity equations (4.2), a number of problems can be anticipated.

First and foremost, a successful modeling of unflatness and distortion effects through polynomial functions requires a well distributed array of points throughout the image. The denser the image point pattern the better the expected fidelity of the empirical correction equations. In off-line industrial photogrammetry an image may contain anywhere from a few tens to thousands of target points. Yet it has been found repeatedly that even in close-range photogrammetric networks possessing strong geometries and dense image point arrays problems of overparameterization arise. The resulting singularities can be detected and removed by correlation and covariance analysis techniques. Such an approach does not always guarantee that overparameterization will not lead to accuracy degradation in the photogrammetric triangulation.

The likelihood of problems arising increases with any reduction in both image point density and object space control. The consequences for machine vision networks, which typically involve a modest number of target points, may be severe. Indeed, experience suggests that in such networks the determination of the coefficients of out-of-plane deformation and in-plane distortion via a direct solution of the collinearity equations should not be contemplated. Having said that, however, there are two terms from (4.11) which are very applicable to CCD cameras and which readily lend themselves to recovery via the photogrammetric self-calibration approach. These are the term which models any differential scaling between the horizontal and vertical pixel spacings, and the term for the non-orthogonality of the x and y axes. When these terms only are considered (4.11) can be reduced to

$$\Delta x_f = a_1 \bar{x} + a_2 \bar{y}$$

$$(4.12)$$

$$\Delta y_f = 0 \; .$$

For the following discussion, the in-plane distortion model is confined to that of (4.12) and the out-of-plane deformation correction is not further considered. Studies involving more comprehensive focal plane correction models for CCD cameras have been conducted [3,4]. The photogrammetric networks involved comprised multiple camera stations and dense object point arrays, characteristics which are rarely exhibited in everyday vision applications.

4.3.7 Interior Orientation Elements

Within the collinearity equation model (4.2), errors in the adopted values of interior orientation elements c, x_0 and y_0 will also give rise to image coordinate perturbations. It is therefore important to consider the calibration of these parameters, which are best isolated through a reformulation of (4.2):

$$x + \left(-x_0 - \frac{\bar{x}}{c} \Delta c \right) + \Delta x = -c_0 \frac{R1}{R3}$$

$$(4.13)$$

$$y + \left(-y_0 - \frac{\bar{y}}{c} \Delta c \right) + \Delta y = -c_0 \frac{R2}{R3}$$

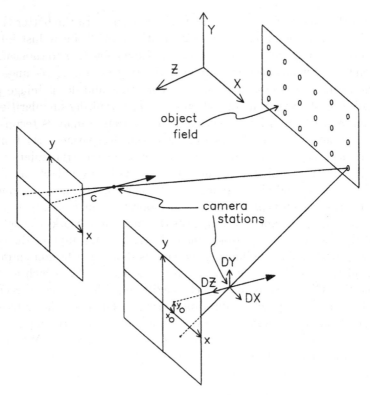

Fig. 4.7. Stereo imaging configuration.

where Δc is the correction to the assumed principal distance c_0. While it is possible to calibrate x_0, y_0 and c in a laboratory utilizing optics techniques [6], the chosen method in analytical photogrammetry is to determine these parameters along with the exterior orientation elements in a general solution of the extended collinearity equations (4.13), through either self-calibration or test-range calibration.

In order to pursue the self-calibration approach it is important to understand the projective coupling between interior and exterior orientation parameters. A simple illustration is provided in Fig. 4.7 where a stereo configuration of two camera stations (either one or two cameras) is shown. The alignment of the object space coordinate system is such that the Z-axis is parallel to the cameras' pointing direction and the X- and x-axes are also parallel. Now, consider the relationship between x_0, y_0 and Δc, and X, Y and Z.

It should first be obvious that with the geometry shown it would not be possible to solve for both the interior and exterior orientation elements via (4.13). The principal point shift is projectively equivalent to a shift of each camera station, say by DX and DY. Moreover, a change in principal distance is projectively equivalent to a shift DZ of the exposure station.

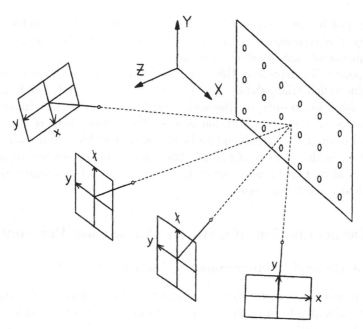

Fig. 4.8. Network to afford self-calibration of interior orientation elements (one camera, four stations).

The coupling between camera station location and principal distance is broken only by introducing scale variation into the image-to-object space transformation. This is best approached through adopting a convergent imaging configuration and/or an object field which is well distributed in three dimensions. A simple diversity of camera roll angle is enough to diminish the correlation between exterior orientation elements and the principal point offset. Figure 4.8 shows a one-camera, four-station network that would facilitate the recovery of interior orientation elements via a general solution of the collinearity equations.

Referring back to Fig. 4.7, it is interesting to consider the impact of biases in the interior orientation values on the photogrammetric triangulation process. If the exterior orientation elements are assumed fixed; that is, the object point is triangulated from two fixed camera stations, then positional errors will result from the presence of biases in x_0, y_0 and c. Moreover, the projection will display an affinity (different scale factors for XY and Z) due to the bias in c.

On the other hand, if the restitution involves computation of exterior orientation elements along with object space coordinates via the collinearity or DLT model, then no biases in object space coordinates will occur due to the projective absorption of errors in interior orientation elements by the exterior orientation parameters. Thus, good results are obtained in spite of the presence of poor calibration. In the convergent geometry of Fig. 4.8 the decoupling of interior and exterior orientation parameters precludes significant projective compensation. Hence, errors will arise in the object space as a consequence of interior orientation biases.

The point to be made from this discussion is that for any photogrammetric or machine vision triangulation system it is difficult to study the impact of system calibration in isolation. Network geometry plays a big part in assessing the influence of camera calibration errors. The danger here is that while successful triangulation results might be obtained with little or no calibration information in the network of Fig. 4.7, there might be a tendency to assume the same will be true for more general, convergent imaging geometries. This will assuredly not be the case.

4.4   Determination of Camera Calibration Parameters

4.4.1   A General Photogrammetric Model

A general functional model for the solution of the collinearity equations is obtained through a linearization of (4.13). In matrix form this model can be expressed as follows:

$$v = A_1 x_1 + A_2 x_2 + A_3 x_3 - l \qquad (4.14)$$

where v is a vector of image coordinate residuals; x_1, x_2 and x_3 are vectors comprising the parameters for the exterior orientation elements, object space coordinates, and interior orientation and additional calibration parameters, respectively; A_1, A_2 and A_3 are the associated design matrices; and l is a vector of discrepancy values. A solution of (4.14) via a linear least-squares approach is the basis of the well-known photogrammetric bundle adjustment with self-calibration.

The step from the functional model of (4.14) to a practical algorithm for the bundle adjustment with self-calibration is a significant one, and one which is beyond the scope of the present discussion. For a more complete account of the subject the reader is referred to the companion paper on system self-calibration by Gruen & Beyer [8] and also to the photogrammetric literature [13]. Here the coverage is kept general in nature; the determination of system calibration parameters via a solution to (4.14) will be discussed without detailed reference to algorithmic aspects.

The vector x_3 comprises the parameters of interior orientation, distortion, and the x/y scale ratio and shear terms of in-plane distortion:

$$x_3 = \begin{pmatrix} x_0 & y_0 & \Delta c & K_1 & K_2 & K_3 & P_1 & P_2 & a_1 & a_2 \end{pmatrix}^\top . \qquad (4.15)$$

The design matrix for a single camera, under the assumption that the additional parameters of (4.15) are invariant from image to image, is given as

$$\boldsymbol{A}_3 = \begin{pmatrix} -1 & 0 & -\bar{x}/c & \bar{x}r^2 & \bar{x}r^4 & \bar{x}r^6 & \left(2\bar{x}^2 + r^2\right) & 2\bar{x}\bar{y} & \bar{x} & \bar{y} \\ 0 & -1 & -\bar{y}/c & \bar{y}r^2 & \bar{y}r^4 & \bar{y}r^6 & 2\bar{x}\bar{y} & \left(2\bar{y}^2 + r^2\right) & 0 & 0 \end{pmatrix} . \quad (4.16)$$

The determinability of the parameters forming $\boldsymbol{x}_3$ is not always assured. On-the-job calibration approaches require that special attention be paid to network geometry, in order to effectively recover the calibration parameters.

4.4.2   Test-Range Calibration

As mentioned, test-range calibration implies that an object space control field of known XYZ coordinates is available for the photogrammetric calibration process. In the presence of known, "fixed" XYZ object point coordinates, (4.14) can be simplified through elimination of the correction terms $\boldsymbol{x}_2$, which by definition will be zero in this case. What remains is the following:

$$\boldsymbol{v} = \boldsymbol{A}_1\boldsymbol{x}_1 + \boldsymbol{A}_3\boldsymbol{x}_3 - \boldsymbol{l} \qquad (4.17)$$

Here, $\boldsymbol{x}_1$ comprises the six elements of exterior orientation and $\boldsymbol{x}_3$ is as per (4.15). The solution of (4.17), again by least-squares, amounts to a self-calibrating camera resection. Test-range calibration can be carried out with a single photograph or with multiple images. In either case the geometry of the control field is important, and a minimal array of eight targets would be necessary to solve for the exterior orientation elements and the 10 parameters comprising $\boldsymbol{x}_3$.

The computational effort involved with the iterative least-squares solution to test-range calibration is reasonably modest, the most intensive requirement in the one-photo case being the inversion of a 16×16 symmetric, positive-definite matrix. In practical applications projective coupling between parameters invariably necessitates the suppression of a subset of both interior orientation and calibration terms to ensure that only statistically significant parameters are recovered. Some general characteristics of test-range calibration which are pertinent to close-range photogrammetric networks, and potentially also to machine vision, are as follows:

(i) To facilitate the recovery of the interior orientation elements x_0, y_0 and c the control point field must be well distributed in three dimensions. Even in such circumstances the projective coupling with the exterior orientation parameters can be very significant.

(ii) The use of multiple photo stations, with varying image scales (but the same focal setting) and camera roll angles greatly enhances the determinability of x_0, y_0 and c.

(iii) As has been previously mentioned, with most nonphotogrammetric lenses (those not designed to display minimal distortion) it is likely that only the K_1 term will be required to fully account for radial distortion effects.

(iv) The decentering distortion parameters P_1 and P_2 typically exhibit a high degree of projective coupling with both x_0 and y_0, and the elements of exterior orientation. The determinability of these parameters is also enhanced by the use of both multiple camera stations and a diversity of camera roll angles. The high correlation between P_1, P_2 and x_0, y_0 can be expected to persist even in these instances. This is because the influences of decentering distortion are akin to those produced by inserting a thin prism in front of the lens, the first-order effect of which is to produce a shift in the principal point. In many practical applications of CCD cameras P_1 and P_2 can be suppressed from x_3.

(v) The greater the image point density the better the expected recovery of the calibration parameters. This statement applies to all the terms comprising x_3 but is most pertinent to the scale ratio a_1, shear term a_2, and the parameters of radial and decentering distortion. Photogrammetrists have a great deal of respect and affection for observational data redundancy. Indeed the fact that analytical restitution models are so overdetermined has considerable bearing on the high measurement reliability of multistation, close-range photogrammetric networks. The downside of this aspect for test-range calibration is the need for suitable, pre-surveyed control point configurations. Portable control "frameworks" have found application in machine vision but their suitability is not universal.

vi) Test-range calibration systems for video cameras can be configured to be autonomous, while still embodying major attributes such as automated data screening and error detection, and statistical evaluation which are found in traditional off-line computation schemes.

4.4.3   Self-Calibration

Provision of a fixed control point array in the test-range calibration approach means that the relative orientation between bundles of rays from different images is not taken into account. Yet relative orientation is a fundamental process of analytical restitution and failure to incorporate this information is a major shortcoming of calibration techniques which employ the resection model. Self-calibration via the full collinearity model (4.14), displays no such shortcomings. This model has the flexibility of incorporating all observational data related both to the image coordinates and the parameters forming x_1, x_2 and x_3, while fully utilizing all geometric relationships embodied in the collinearity equations, both in terms of relative and absolute orientation.

The increased rigor and flexibility of the self-calibration approach comes at a price, and the cost is in the area of computational effort. In a test-range calibration employing a single image the order of the matrices involved reaches 16 at the maximum if the calibration model of (4.15) is employed. A multistation self-calibrating bundle network employing k cameras, m photos and n points will necessitate the solution of an equation system of order $10k + 6m + 3n$. To take an example, consider a network of a single camera, six images and 200 object points. The resulting normal equations of the least-squares bundle adjustment will then be of order 646. The computational considerations involved in handling the bundle adjustment are therefore not trivial.

On the other hand they are far from insurmountable. There are numerous PC-based self-calibration software packages available on the commercial photogrammetry market. While these are all geared for off-line computation, on-line bundle triangulation is both possible and practical in some instances (aerial triangulation, for example). The purpose here, however, is not to dwell too much on the computational aspects and requirements of the self-calibrating bundle adjustment, but more to briefly overview some of its features in terms of camera calibration.

The first point to be made is that by and large the six characteristics listed above for test-range calibration are also pertinent to self-calibration, but some distinctions are worthy of note. Firstly, the bundle adjustment approach does not require that the object point array be well distributed in three dimensions, but where it is not a highly convergent imaging geometry is necessary for a strong recovery of the principal distance. Secondly, the self-calibration approach does not require any a priori knowledge of object point coordinates, though the provision of constrained target coordinates can only improve the recovery of calibration parameters. It is both possible and practical to self-calibrate a camera without the provision of any control, including object space scale. The third distinction to note is that the determinability of camera calibration is greatly enhanced in the self-calibration approach by virtue of the incorporation of the geometric constraints of relative orientation.

Removal of the requirement for an object space control field affords considerable flexibility in the self-calibration method. This can be illustrated by highlighting an autonomous and flexible approach which has been utilized for exterior orientation determination of fixed configurations of multiple CCD cameras. Consider the case of three or four cameras mounted in a "work cell" configuration such that target points can be triangulated throughout a particular working volume. The term "digital coordinate measuring machine" has been used to describe this configuration, which is illustrated in Fig. 4.9. Within this volume a bar with two or more targets is moved to various positions and at each an image is "snapped". The net outcome is the provision of an object space target array of sufficient density to enable the determination of camera orientation and position within an assigned XYZ coordinate system [19]. If point-to-point distances are known on the bar, correct object space scale can also be established. Moreover, the provision of this object space constraint information can enhance the recovery of the camera calibration parameters. The determinability of interior orientation elements should be closely monitored, however, in such a configuration.

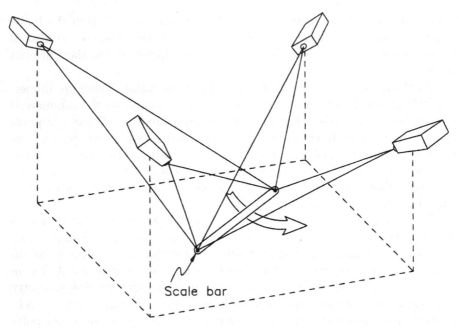

Fig. 4.9. "Work cell" configuration of CCD cameras imaging a scale bar with a target at each end.

4.4.4   Distortion Calibration via the Plumbline Technique

In a distortion free camera all straight lines in object space would project as straight lines on the image plane. Departures of linearity of the imaged lines are then attributable to lens distortion. Consideration of this principle provides the basis for the analytical plumbline technique for the calibration of radial and decentering distortion which was developed by Brown [11]. The equation of an arbitrary straight line on the image can be derived through consideration of Fig. 4.10 as

$$x' \sin \theta + y' \cos \theta - p = 0 \qquad (4.18)$$

where p is the perpendicular distance from the line to the origin of the (x', y') coordinate system, and θ is the angle subtended by the perpendicular and the y'-axis. The necessary correction equations to bring the observed image coordinates (x, y) onto the straight line are then formed as

$$x' = \bar{x} + \Delta x_r + \Delta x_d$$

$$\qquad (4.19)$$

$$y' = \bar{y} + \Delta y_r + \Delta y_d$$

where the correction models for radial and decentering distortion are those of (4.5) and (4.8), respectively. Through a combination of (4.18) and (4.19) a nonlinear functional model for the plumbline technique is obtained. This can be presented in the form:

$$f\left(x, y, x_0, y_0, K_1, K_2, K_3, P_1, P_2, \theta, p\right) = 0 \qquad (4.20)$$

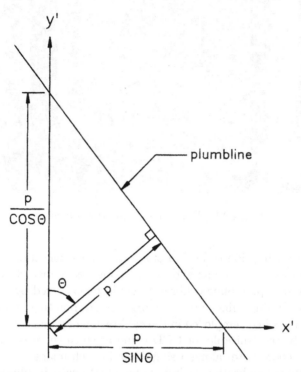

Fig. 4.10. Geometry for analytical plumbline method.

Although the principal point offsets x_0 and y_0 appear in this model they are rarely recoverable. What is determined from the model of (4.20) are the radial lens distortion coefficients K_1, K_2 and K_3, and the decentering distortion parameters P_1 and P_2.

The effectiveness of the plumbline technique is largely dependent on the number and distribution of the straight line images available. Figure 4.11 shows a plumbline range utilized for the calibration of distortion in large- and medium-format metric film cameras. The figure largely explains the origin of the name "plumbline calibration". Plumblines per se are not required; any suitable set of straight lines will suffice. Since the basic correction terms, (4.19), contain both x and y image coordinates it is desirable to have the plumbline images in more than one orientation. A practical way to achieve this is to take one photo of a range such as that shown in Fig. 4.11, and then to expose a second image with the camera rolled about its axis by 90°.

Fig. 4.11. Plumbline calibration range.

Under the assumption of both a stable interior orientation, and the absence of photo-variant in-plane and out-of-plane image distortion, the set of data required for a plumbline calibration could be gathered by taking multiple exposures of a single line, with changing camera pointing and orientation. This approach has been employed in the calibration of aerial cameras [20] and it is potentially very suitable for CCD cameras since there is only the need to extract the coordinates along a single line in each image.

The plumbline calibration technique was little used in photogrammetry until the mid 1980s when automated image mensuration removed the need for tedious manual measurement of point coordinates along the imaged lines. Today, the technique is enjoying wider usage in close-range photogrammetry [21], including the area of digital photogrammetry [22,6,23] where it can be highly automated. The distortion parameters determined in a plumbline calibration can also be employed as constraints in a subsequent test range or self-calibration to reduce the influence of projective coupling between interior and exterior orientation elements, especially in circumstances where the network geometry is a little weak.

4.5   Concluding Remarks

In examining some of the "calibration" procedures employed in computer and machine vision it is apparent that a different philosophy prevails in cases to that found in photogrammetry. Vision systems are configured to measure within the object space. Thus, it is not surprising that the calibration operation in vision systems often concentrates on a modeling of object space triangulation errors such that positional "correction" functions are generated and applied to triangulated XYZ coordinates. This approach, which obviously relies on the provision of object space "control", does not expressly consider perturbations to collinearity due to the cameras employed.

Photogrammetry, on the other hand, concentrates on achieving calibration through comprehensive mathematical modeling within image space. Departures from collinearity are modeled in terms of image space parameters. The difference in these two approaches is not trivial. From a practical point of view the chief distinction is that whereas the former approach is camera-configuration specific, the latter is applicable to any camera station network in which the calibrated cameras are employed.

As a final note on analytical photogrammetric calibration, it is again stressed that these techniques are most suited to the pre-calibration of CCD cameras prior to their utilization in a time-constrained vision environment. Subsequent instabilities in calibration parameters are unlikely to be of much practical consequence in most instances.

Modern analytical photogrammetry offers a comprehensive camera system calibration technique which is versatile, efficient and can accommodate off-the-shelf CCD cameras and lenses. Moreover, photogrammetric calibration techniques produce optimal accuracy and can be automated to a significant degree. For numerous vision applications the approach adequately fulfills the two principal calibration purposes mentioned by Tsai [24]: Inferring 3D information from computer image coordinates, and inferring 2D computer image coordinates from 3D information.

References

1. C.S. Fraser. Photogrammetric Measurement to One Part in a Million. Photogrammetric Engineering & Remote Sensing **58(3)**, 305–310, 1992
2. P.G. Gustafson. An Accuracy/Repeatability Test for a Video Photogrammetric Measurement. *Proc. Industrial Vision Metrology*, Winnipeg, July 11–12, SPIE vol. 1526, pp. 36-41, 1991
3. M.R. Shortis, A.W. Burner, W.L. Snow, W.K. Goad. Calibration Tests of Industrial and Scientific CCD Cameras. *Proc. 1st Australian Photogrammetry Conference, Sydney*, Nov. 7–9, Paper 6, 11 p., 1991
4. H.A. Beyer. Evaluating the Geometric Performance of Signal Transmission. *Proc. 1st Australian Photogrammetry Conference*, Sydney, November 7–9, Paper 15, 6 p., 1991
5. W. Boesemann, R. Godding, W. Riechmann. Photogrammetric Investigation of CCD Cameras. *Proc. Close-Range Photogrammetry Meets Machine Vision, ISPRS Comm. V*, Zurich, SPIE vol. 1395, pp. 119–126, 1990
6. A.W. Burner, W.L. Snow, M.R. Shortis, W.K. Goad. Laboratory Calibration and Characterization of Video Cameras. *Proc. Close-Range Photogrammetry meets Machine Vision, ISPRS Comm. V*, Zurich, SPIE vol. 1395, pp. 664–671, 1990

120   Clive S. Fraser

7. R.Y. Tsai. An Efficient and Accurate Camera Calibration Technique for 3D Machine Vision. *Proc. IEEE Computer Vision & Pattern Recognition Conference*, pp. 364–374, Miami Beach, USA 1986
8. A. Gruen, H.A. Beyer. System Calibration through Self-Calibration. *Presented to Workshop on Calibration and Orientation of Cameras in Computer Vision.* ISPRS XVII Congress, Washington, DC, August 2–14, 1992 (Also Chap. 6 of this volume)
9. Y.I. Abdel-Aziz, H.M. Karara. Direct Linear Transformation from Comparator Coordinates into Object Space Coordinates in Close-Range Photogrammetry. *Proc. ASP/UI Symposium on Close-Range Photogrammetry*, pp. 1–18, Urbana, Illinois 1971
10. H. Bopp, H. Krauss. Extension of the 11-Parameter Solution for on-the-job Calibrations of Non-Metric Cameras. International Archives of Photogrammetry **22(5)**, 7–11, 1978
11. D.C. Brown. Close-Range Camera Calibration. Photogrammetric Engineering **37(8)**, 855–866, 1971
12. C.S. Fraser, M.R. Shortis. Variation of Distortion within the Photographic Field. Photogrammetric Engineering & Remote Sensing, **58(6)**, 851–855, 1992
13. H.M. Karara, (Ed.). *Non-Topographic Photogrammetry*, 2nd Ed. American Society of Photogrammetry & Remote Sensing, Falls Church, 445 pages, Virginia 1989
14. D.C. Brown. Decentering Distortion of Lenses. Photogrammetric Engineering **32(3)**, 444–462, 1966
15. D.C. Brown. A Large Format Microprocessor Controlled Film Camera Optimized for Industrial Photogrammetry. *Presented Paper, XV Congress of Photogrammetry & Remote Sensing, Commission V*, Rio de Janeiro, 29 p., 1984
16. H.A. Beyer. Linejitter and Geometric Calibration of CCD Cameras. ISPRS Journal of Photogrammetry & Remote Sensing **45**, 17–32, 1990
17. J. Daehler, Problems in Digital Image Acquisition with CCD Cameras. *Proc. ISPRS Inter Commission Conference on Fast Processing of Photogrammetric Data*, Interlaken, Switzerland, pp. 48–59, June 2–4, 1987
18. R.K. Lenz. Lens Distortion Corrected CCD Camera Calibration with Co-Planar Calibration Points for Real-Time 3D Measurements. *Proc. ISPRS Inter Commission Conference on Fast Processing of Photogrammetric Data*, Interlaken, Switzerland, pp. 60–67, June 2–4, 1987
19. H. Haggrén, J. Heikkilä. Calibration of Close-Range Photogrammetric Stations Using a Free Network Bundle Adjustment. Photogrammetric Journal of Finland, **11(2)**, 21–31, 1989
20. J.G. Fryer, & D.G. Goodin. In-Flight Aerial Camera Calibration from Photography of Linear Features. Photogrammetric Engineering & Remote Sensing, **55(12)**, 1751–1754, 1989
21. J.G. Fryer, D.C. Brown. Lens Distortion for Close-Range Photogrammetry. Photogrammetric Engineering & Remote Sensing, **52(4)**, 51–58, 1986
22. H.A. Beyer. Some Aspects of the Geometric Calibration of CCD Cameras. *Proc. ISPRS Inter Comm. Symp. on Fast Processing of Photogrammetric Data*, Interlaken, Switzerland, June 2–4, pp. 68–81, 1987
23. J.G. Fryer, S.O. Mason. Rapid Lens Calibration of a Video Camera. Photogrammetric Engineering & Remote Sensing **55(4)**, 437–442, 1989

24. A.W. Burner, W.L. Snow, W.K. Goad. Close-Range Photogrammetry with Video Cameras. *Proc. 51st Annual Technical Meeting, American Society of Photogrammetry & Remote Sensing*, pp. 62–77, Washington, 1985

5 Least-Squares Camera Calibration Including Lens Distortion and Automatic Editing of Calibration Points

Donald B. Gennery

Summary

A method is described for calibrating cameras including radial lens distortion, by using known points such as those measured from a calibration fixture. The distortion terms are relative to the optical axis, which is included in the model so that it does not have to be orthogonal to the image sensor plane. A priori standard deviations can be used to apply weight to zero values for the distortion terms and to zero difference between the optical axis and the perpendicular to the sensor plane, so that the solution for these is well determined when there is insufficient information in the calibration data. The initial approximations needed for the nonlinear least-squares adjustment are obtained in a simple manner from the calibration data and other known information. Outliers among the calibration points are removed by means of automatic editing based on analysis of the residuals. The use of the camera model also is described, including partial derivatives for propagating both from object space to image space and vice versa.

5.1 Introduction

The basic camera model that we have been using in the JPL robotics laboratory was originally developed here by Yakimovsky and Cunningham [1]. It included a central perspective projection and an arbitrary affine transformation in the image plane, but it did not include lens distortion. In 1986, the present author developed a better method of calibrating that model, by using a rigorous least-squares adjustment [2]. In 1990, that camera model and the method for its calibration were extended to include radial lens distortion [3]. Further improvements were made in 1992. This chapter describes the improved model, the adjustment algorithm, and the mathematics for the use of the camera model. However, the method of measuring the calibration data (finding the dots in images of a calibration fixture [2]) has not changed, and thus its description will not be repeated here.

The camera model used here should be adequate for producing accurate geometric data, except in the following three situations. First, a fish-eye lens

Springer Series in Information Sciences, Vol. 34
Calibration and Orientation of Cameras in Computer Vision
Eds.: Gruen, Huang © Springer-Verlag Berlin Heidelberg 2001

has a very large distortion for which the distortion polynomial used here would not converge. (In fact, the distortion as defined here can be infinite, since the field of view can exceed 180°). For such a lens the image coordinate should be represented as being ideally proportional to the off-axis angle, instead of the tangent of this angle as in the perspective projection. Then, a small distortion could be added on top of this. Furthermore, the position of the entrance pupil of a fish-eye lens varies greatly with the off-axis angle to the object; therefore, this variation would have to be modeled unless all viewed objects are very far away. Second, even for an ordinary lens, the entrance pupil moves slightly, so that if objects very close to the lens are viewed, the variation again would have to be included in the camera model. Third, if there is appreciable nonradial distortion, such as might be produced by distortion in the image sensor itself or by a lens with badly misaligned elements, it would require a more elaborate model than the radial distortion used here. However, this situation should not occur with a CCD camera with a well-made lens, unless an anamorphic lens is used. (A small amount of decentering in the lens and a CCD synchronization error that is a linear function of the position in the image are subsumed by terms included in the camera model.)

The calibration method presented here applies at only one lens setting. For a zoom lens, a separate calibration would have to be done at each zoom setting. Similarly, if focus is changed, a separate calibration is needed at each focus setting.

In this chapter, for physical vectors in three-dimensional space, the dot product will be indicated by $\cdot$ and the cross product by $\times$. For any such vector v, its length ($\sqrt{v \cdot v}$) will be represented by $|v|$, and the unit vector in its direction ($v/|v|$) will be represented by $\mathrm{unit}(v)$. The derivative of a vector relative to a scalar will be considered to be a vector, the derivative of a scalar relative to a vector will be considered to be a row vector, and the derivative of a vector relative to a vector will be considered to be a matrix.

5.2   Definition of Camera Model

5.2.1   Camera Model Without Distortion

The old camera model [1] (without distortion) consisted of the four 3-vectors c, a, h and v, expressed in object (world) coordinates, which have the following meanings. The entrance pupil point of the camera lens is at position c. Let a perpendicular be dropped from the exit pupil point to the image sensor plane, and let its point of intersection with this plane be denoted by x_c and y_c in image coordinates. (Decentering in the lens causes the actual values of x_c and y_c to differ from those according this definition.) Then a is a unit vector parallel to this perpendicular and pointing outwards from the camera to the scene. Furthermore, let v' and h' be vectors in the sensor plane and perpendicular to the x and y image axes, respectively (usually considered to be horizontal and vertical, respectively, but not necessarily orthogonal) of the image coordinate system, with each magnitude equal to the change in the image coordinate (usually measured in pixels) caused by a change of unity in the tangent of the angle from the $\boldsymbol{a}$ vector to the viewed point, subtended at the entrance pupil point. (If the entrance and exit pupil points coincide with the first and second nodal points, as is approximately the case for typical cameras with other than zoom or telephoto lenses, this is equivalent to saying that the magnitude of $\boldsymbol{h}'$ or $\boldsymbol{v}'$ is equal to the distance from the second nodal point to the sensor plane, expressed in horizontal pixels or vertical pixels, respectively.) Then $\boldsymbol{h} = \boldsymbol{h}' + x_c\boldsymbol{a}$ and $\boldsymbol{v} = \boldsymbol{v}' + y_c\boldsymbol{a}$. Although these definitions may seem rather peculiar, they result in convenient expressions for the image coordinates in terms of the position $\boldsymbol{p}$ of a point in three-dimensional object space, as follows:

$$x = \frac{(\boldsymbol{p} - \boldsymbol{c}) \cdot \boldsymbol{h}}{(\boldsymbol{p} - \boldsymbol{c}) \cdot \boldsymbol{a}} \tag{5.1}$$

$$y = \frac{(\boldsymbol{p} - \boldsymbol{c}) \cdot \boldsymbol{v}}{(\boldsymbol{p} - \boldsymbol{c}) \cdot \boldsymbol{a}} . \tag{5.2}$$

The above equations are given here merely to show the mathematical relationship represented by the camera model excluding distortion. In the equations used below for actual data, the subscript i will be used on quantities associated with individual measured points.

The position $\boldsymbol{c}$ used above is often referred to as the perspective center. Sometimes, this is assumed to be the first nodal point of the lens (the same as the first principal point if the medium is the same on both sides of the lens). However, the rays from the object (extended as straight lines) must pass through the entrance pupil if they are to reach the image, so the center of the entrance pupil is the position from which the camera seems to view the world. Similarly, the rays to the image seem to emanate from the exit pupil. If all viewed objects are at the same distance and the camera is perfectly focused, there is no detectable difference between using nodal points instead of pupil points in the above definitions, but in general the distinction should be made; and, if all calibration points are at the same distance, the calibration cannot determine $\boldsymbol{c}$ anyway. Of course, mathematically the camera model is defined by the equations, so the physical meaning of the parameters does not matter for calibration purposes, as long as they are able to capture the degrees of freedom in the physical situation, and as long as the same equations are used in the camera model adjustment and in the use of the camera model. (For definitions of the terms used here, see any optics textbook, for example [4]).

5.2.2   Inclusion of Distortion

Because of symmetry, the dispacement due to radial distortion is a polynomial (in the distance from the optical axis) containing only odd-order terms. The first-order term is subsumed by the scale factors included in $\boldsymbol{h}$ and $\boldsymbol{v}$, if $\boldsymbol{a}$

is along the optical axis of the lens or if the pupil points coincide with the respective nodal points, but it must be included in the general case.

The distortion polynomial must be defined relative to the optical axis of the lens, in order for the distortion to be radial. If the is perpendicular to the optical axis, this is the a vector. However, to allow for the possibility that they are not exactly perpendicular, a separate unit vector o is used here for the optical axis. Note that if there is no distortion and the pupil points coincide with the nodal points, it is impossible for the calibration to determine the optical axis. (A central projection is equivalent to using a pinhole camera, and a pinhole has no axis.) Therefore, some a priori weight will be applied here to tend to make the o vector equal to the a vector, so that it will be well determined when there is not much distortion. However, when there is a large distortion, the data will outweigh this a priori information, and the two vectors can differ.

Let p_i be the three-dimensional position of any point being viewed. Then the component parallel to the optical axis of the distance to the point is

$$\zeta_i = (p_i - c) \cdot o . \tag{5.3}$$

The vector to the point orthogonally from the optical axis is

$$\lambda_i = p_i - c - \zeta_i o . \tag{5.4}$$

The square of the tangent of the angle from the optical axis to the point (subtended at the entrance pupil point) is

$$\tau_i = \frac{\lambda_i \cdot \lambda_i}{\zeta_i^2} . \tag{5.5}$$

We can define a proportionality coefficient μ_i that produces the amount of distortion when multiplied by the off-axis distance to the point, as a polynomial containing only even-order terms (since we have factored out the first power). Since τ_i is proportional to the square of this off-axis distance, this polynomial can be written as follows:

$$\mu_i = \rho_0 + \rho_1 \tau_i + \rho_2 \tau_i^2 + \dots \tag{5.6}$$

where the ρ's are the distortion coefficients. We use only coefficients up to ρ_2, since higher-order terms are negligible except for wide-angle lenses. (In fact, often ρ_2 is negligible.) Some a priori weight (usually a small amount) will be applied to tend to make the ρ's equal to zero, so that they will be well determined when there is not enough information in the calibration images. (This is especially important for high-order coefficients when there are not many calibration points, and for ρ_0 if o and a are nearly equal or if the pupil points nearly coincide with the nodal points.) Note that ρ_0 does not actually represent radial distortion in the usual sense, but is merely a scale factor in the plane orthogonal to o, whereas the scale factors included in h and v are in a plane orthogonal to a.

The effect of distortion can now be written as follows:

$$p'_i = p_i + \mu_i \lambda_i \tag{5.7}$$

where p_i is the true position of the point and p'_i is its apparent position because of distortion.

Then the distorted point can be projected into image coordinates (denoted by $\hat{x}_i$ and $\hat{y}_i$) by using p'_i instead of p in (5.1, 5.2). However, for efficiency these equations are expressed in terms of the intermediate quantities α_i, β_i, and γ_i, as follows, since these quantities will be needed again in later sections:

$$\alpha_i = (p'_i - c) \cdot a \tag{5.8}$$

$$\beta_i = (p'_i - c) \cdot h \tag{5.9}$$

$$\gamma_i = (p'_i - c) \cdot v \tag{5.10}$$

$$\hat{x}_i = \frac{\beta_i}{\alpha_i} \tag{5.11}$$

$$\hat{y}_i = \frac{\gamma_i}{\alpha_i} . \tag{5.12}$$

(The reason for using the circumflex over x and y is to represent computed values, to distinguish them from measured values that will be used in Sect. 5.4.)

Therefore, the complete camera model consists of the five vectors c, a, h, v, and o, where a and o are unit vectors, and the distortion coefficients ρ_0, ρ_1, and ρ_2. This is 18 parameters in all, of which two are redundant because of the unit vectors. (The computations can be extended in a straightforward way to include higher-order ρ's, if they are ever needed.)

5.3 Partial Derivatives

Partial derivatives of several quantities defined in Sect. 5.2 will be needed in doing the least-squares adjustment of the camera model and in propagating error estimates when using the camera model. These are as follows:

$$\frac{\partial \lambda_i}{\partial p_i} = I - oo^{\mathrm{T}} \tag{5.13}$$

$$\frac{\partial \lambda_i}{\partial o} = -\zeta_i I - o(p_i - c)^{\mathrm{T}} \tag{5.14}$$

$$\frac{\partial \mu_i}{\partial \tau_i} = \rho_1 + 2\rho_2 \tau_i \tag{5.15}$$

$$\frac{\partial p_i'}{\partial o} = \frac{\partial \mu_i}{\partial \tau_i} \lambda_i \left(\frac{2}{\zeta_i^2} \lambda_i^{\mathrm{T}} \frac{\partial \lambda_i}{\partial o} - \frac{2\tau_i}{\zeta_i}(p_i - c)^{\mathrm{T}} \right) + \mu_i \frac{\partial \lambda_i}{\partial o} \tag{5.16}$$

$$\frac{\partial p_i'}{\partial p_i} = I + \frac{\partial \mu_i}{\partial \tau_i} \lambda_i \left(\frac{2}{\zeta_i^2} \lambda_i^{\mathrm{T}} \frac{\partial \lambda_i}{\partial p_i} - \frac{2\tau_i}{\zeta_i} o^{\mathrm{T}} \right) + \mu_i \frac{\partial \lambda_i}{\partial p_i} \tag{5.17}$$

$$\frac{\partial \hat{x}_i}{\partial p_i'} = \frac{h^{\mathrm{T}}}{\alpha_i} - \frac{\beta_i a^{\mathrm{T}}}{\alpha_i^2} = \frac{h^{\mathrm{T}} - \hat{x}_i a^{\mathrm{T}}}{\alpha_i} \tag{5.18}$$

$$\frac{\partial \hat{y}_i}{\partial p_i'} = \frac{v^{\mathrm{T}}}{\alpha_i} - \frac{\gamma_i a^{\mathrm{T}}}{\alpha_i^2} = \frac{v^{\mathrm{T}} - \hat{y}_i a^{\mathrm{T}}}{\alpha_i} \tag{5.19}$$

$$\frac{\partial \hat{x}_i}{\partial a} = -\frac{\beta_i(p_i'-c)^{\mathrm{T}}}{\alpha_i^2} = -\frac{\hat{x}_i(p_i'-c)^{\mathrm{T}}}{\alpha_i} \tag{5.20}$$

$$\frac{\partial \hat{y}_i}{\partial a} = -\frac{\gamma_i(p_i'-c)^{\mathrm{T}}}{\alpha_i^2} = -\frac{\hat{y}_i(p_i'-c)^{\mathrm{T}}}{\alpha_i} \tag{5.21}$$

where I is the 3×3 identity matrix. The first three of these (derivatives of λ_i and μ_i) are only intermediate quantities needed in obtaining the others and will not be used elsewhere.

5.4   Adjustment of Camera Model

5.4.1   Data for Adjustment

The calibration data consists of a set of points, with for each point i its three-dimensional position p_i in object coordinates and its measured two-dimensional position x_i and y_i in image coordinates. Given information consists of the following: the a priori standard deviation σ_d (in radians) of the difference between the optical axis o and the a vector, the a priori standard deviations about zero of each distortion coefficient σ_0, σ_1 and σ_2 (dimensionless quantities that in effect are proportionality constants when at an angle of $45°$ from the optical axis), the minimum standard deviation of measured image-coordinate positions σ_m, the nominal focal length of the camera f, the nominal horizontal and vertical pixel spacings p_h and p_v (in the same units as f), the number of columns s_h and rows s_v of pixels in the camera, and the approximate position of the camera c_o in object coordinates. (Reasonable values for the standard deviations are $\sigma_d = 0.01, \sigma_0 = 1$ for zoom and telephoto lenses or 0.1 for ordinary lenses, $\sigma_1 = 1$, and $\sigma_2 = 1$. On the other hand, any of the $\rho's$ could be forced to be zero, if desired, by making their standard deviations very small, perhaps 10^{-5}. The quantities $f, p_h, p_v, s_h, s_v,$ and c_o are used primarily in obtaining an initial approximation for iterating, and thus their exact values are not important.) The desired result consists of the camera model parameters defined in Sect. 5.2, which where convenient will be assembled into the 18-vector $g = \left[c^{\mathrm{T}} a^{\mathrm{T}} h^{\mathrm{T}} v^{\mathrm{T}} o^{\mathrm{T}} \rho_0 \rho_1 \rho_2 \right]^{\mathrm{T}}$, and their 18×18 covariance matrix C_{gg}, which indicates the accuracy of the result.

In order to obtain a solution, the input points must not be all coplanar, and they must be distributed over the image space. Since there are 16 independent parameters in the camera model and each point contains two dimensions of information in the image plane, at least eight points are needed, or six if the a priori standard deviations are small enough to add appreciable information. However, it is highly desirable to have considerably more points than this minimum in order obtain an accurate solution for all of the parameters.

5.4.2 Initialization

In order to obtain initial values for iterating, first a point close to the camera axis is found by selecting p_a to be the p_i for which x and y are closest to $s_h/2$ and $s_v/2$, respectively. Then this and the given data are used to compute the initial approximations to the camera model vectors, as follows:

$$c_o \text{ is given} \tag{5.22}$$

$$a_o = \text{unit}(p_a - c_o) \tag{5.23}$$

$$h_o = \frac{f}{p_h} \text{unit}(a_o \times u) + \frac{s_h}{2} a_o \tag{5.24}$$

$$v_o = \frac{f}{p_v} \text{unit}(a_o \times h_o) + \frac{s_v}{2} a_o \tag{5.25}$$

$$o_o = a_o \tag{5.26}$$

where u is a vector pointing upwards in object space, and where it is assumed that the image x axis points to the right and the image y axis points down (from the upper left corner of the image). (If the y axis points up in the image, the sign of v_o should be reversed.) The initial values for ρ_0, ρ_1, and ρ_2 are zero.

The a priori weight matrix is computed as follows:

$$
N_o = \begin{bmatrix}
0 & 0 & 0 & 0 & 0 & 0 & 0 & 0 & 0 & 0 & 0 & 0 & 0 & 0 & 0 & 0 & 0 & 0 \\
0 & 0 & 0 & 0 & 0 & 0 & 0 & 0 & 0 & 0 & 0 & 0 & 0 & 0 & 0 & 0 & 0 & 0 \\
0 & 0 & 0 & 0 & 0 & 0 & 0 & 0 & 0 & 0 & 0 & 0 & 0 & 0 & 0 & 0 & 0 & 0 \\
0 & 0 & 0 & \frac{1}{\sigma_d^2} & 0 & 0 & 0 & 0 & 0 & 0 & 0 & 0 & -\frac{1}{\sigma_d^2} & 0 & 0 & 0 & 0 & 0 \\
0 & 0 & 0 & 0 & \frac{1}{\sigma_d^2} & 0 & 0 & 0 & 0 & 0 & 0 & 0 & 0 & -\frac{1}{\sigma_d^2} & 0 & 0 & 0 & 0 \\
0 & 0 & 0 & 0 & 0 & \frac{1}{\sigma_d^2} & 0 & 0 & 0 & 0 & 0 & 0 & 0 & 0 & -\frac{1}{\sigma_d^2} & 0 & 0 & 0 \\
0 & 0 & 0 & 0 & 0 & 0 & 0 & 0 & 0 & 0 & 0 & 0 & 0 & 0 & 0 & 0 & 0 & 0 \\
0 & 0 & 0 & 0 & 0 & 0 & 0 & 0 & 0 & 0 & 0 & 0 & 0 & 0 & 0 & 0 & 0 & 0 \\
0 & 0 & 0 & 0 & 0 & 0 & 0 & 0 & 0 & 0 & 0 & 0 & 0 & 0 & 0 & 0 & 0 & 0 \\
0 & 0 & 0 & 0 & 0 & 0 & 0 & 0 & 0 & 0 & 0 & 0 & 0 & 0 & 0 & 0 & 0 & 0 \\
0 & 0 & 0 & 0 & 0 & 0 & 0 & 0 & 0 & 0 & 0 & 0 & 0 & 0 & 0 & 0 & 0 & 0 \\
0 & 0 & 0 & 0 & 0 & 0 & 0 & 0 & 0 & 0 & 0 & 0 & 0 & 0 & 0 & 0 & 0 & 0 \\
0 & 0 & 0 & -\frac{1}{\sigma_d^2} & 0 & 0 & 0 & 0 & 0 & 0 & 0 & 0 & \frac{1}{\sigma_d^2} & 0 & 0 & 0 & 0 & 0 \\
0 & 0 & 0 & 0 & -\frac{1}{\sigma_d^2} & 0 & 0 & 0 & 0 & 0 & 0 & 0 & 0 & \frac{1}{\sigma_d^2} & 0 & 0 & 0 & 0 \\
0 & 0 & 0 & 0 & 0 & -\frac{1}{\sigma_d^2} & 0 & 0 & 0 & 0 & 0 & 0 & 0 & 0 & \frac{1}{\sigma_d^2} & 0 & 0 & 0 \\
0 & 0 & 0 & 0 & 0 & 0 & 0 & 0 & 0 & 0 & 0 & 0 & 0 & 0 & 0 & \frac{1}{\sigma_0^2} & 0 & 0 \\
0 & 0 & 0 & 0 & 0 & 0 & 0 & 0 & 0 & 0 & 0 & 0 & 0 & 0 & 0 & 0 & \frac{1}{\sigma_1^2} & 0 \\
0 & 0 & 0 & 0 & 0 & 0 & 0 & 0 & 0 & 0 & 0 & 0 & 0 & 0 & 0 & 0 & 0 & \frac{1}{\sigma_2^2}
\end{bmatrix}
$$

$$(5.27)$$

where the fact that the off-diagonal terms for a and o are the negative of the main-diagonal terms causes the standard deviation σ_d to apply to the difference of a and o. Additional weight can be applied to any other of the initial approximation values, by adding the reciprocal of the variance to the appropriate main diagonal element of N_o. For example, the position of the camera may be known sufficiently well to enter this information into the solution by adding appropriate weight in the first three diagonal elements of N_o.

5.4.3   Iterative Solution

The method described here does a rigorous least-squares adjustment [5] in which the camera model parameters are adjusted to minimize the sum of the squares of the residuals (differences between measured and adjusted positions of points) in the image plane. Since the problem is nonlinear, this requires an iterative solution. However, the method converges rapidly unless far from the correct solution, and a reasonably good approximation to start the iterations was obtained in Sect. 5.4.2.

The program has an inner loop for iterating the nonlinear solution and an outer loop for editing (removing erroneous points) by a previously developed general method [6,7]. The steps in these computations for the problem here are as follows.

1. (This is the beginning of the edit loop.) Set c, a, h, v, and o to their initial approximations $(c_o, a_o, h_o, v_o,$ and $o_o)$, set ρ_0, ρ_1 and ρ_2 to zero, and set the estimated measurement variance σ^2 to 1, as an initial approximation.

2. (This is the beginning of the iteration loop.) First, the 2×18 matrix of partial derivatives of the constraints (unit$(a) = 1$ and unit$(o) = 1$) relative to g (the parameters) is

$$K = \begin{bmatrix} 0 & 0 & 0 & \text{unit}(a)^T & 0 & 0 & 0 & 0 & 0 & 0 & 0 & 0 & 0 & 0 & 0 & 0 \\ 0 & 0 & 0 & 0 & 0 & 0 & 0 & 0 & 0 & 0 & 0 & 0 & \text{unit}(o)^T & 0 & 0 & 0 \end{bmatrix}.$$
(5.28)

For each point i currently retained, compute $\hat{x}_i, \hat{y}_i$ and the preliminary quantities by the equations in Sect. 5.2.2 and compute the partial derivatives of $p'_i, \hat{x}_i,$ and $\hat{y}_i$ as in Sect. 5.3. Then the 2×18 matrix of partial derivatives of $\hat{x}_i,$ and $\hat{y}_i$ relative to g is

$$A_i = \begin{bmatrix} -\frac{\partial \hat{x}_i}{\partial p'_i}\frac{\partial p'_i}{\partial p_i} & \frac{\partial \hat{x}_i}{\partial a} & \frac{(p'_i - c)^T}{\alpha_i} & 0^T & \frac{\partial \hat{x}_i}{\partial p'_i}\frac{\partial p'_i}{\partial o} & \frac{\partial \hat{x}_i}{\partial p'_i}\lambda_i & \tau_i\frac{\partial \hat{x}_i}{\partial p'_i}\lambda_i & \tau_i^2\frac{\partial \hat{x}_i}{\partial p'_i}\lambda_i \\ -\frac{\partial \hat{y}_i}{\partial p'_i}\frac{\partial p'_i}{\partial p_i} & \frac{\partial \hat{y}_i}{\partial a} & 0^T & \frac{(p'_i - c)^T}{\alpha_i} & \frac{\partial \hat{y}_i}{\partial p'_i}\frac{\partial p'_i}{\partial o} & \frac{\partial \hat{y}_i}{\partial p'_i}\lambda_i & \tau_i\frac{\partial \hat{y}_i}{\partial p'_i}\lambda_i & \tau_i^2\frac{\partial \hat{y}_i}{\partial p'_i}\lambda_i \end{bmatrix}$$
(5.29)

(since $\partial \hat{x}_i / \partial c = -\partial \hat{x}_i / \partial p_i$, and similarly for $\hat{y}_i$), and the discrepancies between measured and computed data are

$$e_i = \begin{bmatrix} x_i - \hat{x}_i \\ y_i - \hat{y}_i \end{bmatrix}.$$
(5.30)

Compute the matrix of coefficients in the normal equations N (18×18), the "constants" in the normal equations t (18×1), and the sum of the squares of the discrepancies (which become the residuals when the solution has converged) q as follows:

$$N = N_o + \frac{nf^2}{\sigma^2 p_h p_v} K^T K + \frac{1}{\sigma^2} \sum_i A_i^T A$$
(5.31)

$$t = N_o(g_o - g) + \frac{nf^2}{\sigma^2 p_h p_v} K^T \begin{bmatrix} 1 - |a| \\ 1 - |o| \end{bmatrix} + \frac{1}{\sigma^2} \sum_i A_i^T e_i$$
(5.32)

$$q = \sum_i e_i^T e_i$$
(5.33)

where the summations are over all points currently retained, n is the total number of these points, g represents the current parameter values, and g_o represents the a priori parameter values (initial approximations). The first term in (5.31) and (5.32) applies the a priori weight to the initial values, and the second term applies some weight to the constraints. These constraint terms mathematically have no effect on the solution, since the exact constraints are applied below in steps 3 and 5. However, they are necessary to prevent the solution without the constraints from being singular, and that will be computed first (in step 3, as $N^{-1}t$) before the constraint equation is applied. The scale factor chosen for these terms above causes them to have about the same magnitude as the result of the main summation for N, so that numerical accuracy is preserved.

3. Compute the following (using the exact general constraint equations provided by Mikhail [5]):

$$M = KN^{-1}K^{\mathrm{T}} \tag{5.34}$$

$$d = N^{-1}t + N^{-1}K^T M^{-1}\left(\begin{bmatrix} 1 - |a| \\ 1 - |o| \end{bmatrix} - KN^{-1}t\right). \tag{5.35}$$

Note that a and o are unit vectors on the first iteration (because of the initial approximations) and on the last iteration (within the convergence tolerance), but on intermediate iterations they in general are not. The comparison of their magnitudes with unity in the above equation is what causes them to converge to unit vectors. Then add d to the current values of the camera model parameters to produce the new values, as follows:

$$g \leftarrow g + d \tag{5.36}$$

The estimate of measurement variance is

$$\sigma^2 = \max\left(\frac{q}{2n - 14}, \sigma_m^2\right) \tag{5.37}$$

where n is the number of points currently retained (not rejected), that is, the number of points actually used to obtain q. The denominator in (5.37) would be $2n - 16$ if there were no a priori weights. The value $2n - 14$ is an approximation, based on the fact that there is usually a large a priori weight forcing o to be nearly equal to a, but not much weight forcing ρ_0, ρ_1, and ρ_2 to be nearly zero. The exact value is not very important, since the number of measurements $2n$ usually is much larger than 14, and usually not much accuracy is needed or is attainable with variances, anyway.

4. If the magnitude (length of vector) of the change in c (first three elements of d) is less than $10^{-5}|p_a - c_o|$, the magnitude of the change in a (second three elements of d) is less than 10^{-5}, the magnitude of the change in h (third three elements of d) is less than $10^{-5}f/p_h$, the magnitude of the change in v (fourth three elements of d) is less than $10^{-5}f/p_v$, the magnitude of the change in o (fifth three elements of d) is less than 10^{-5}, and the absolute values of the changes in the ρ's (the last three elements of d) are each less than 10^{-3}, then go to step 5 (exit from the iteration loop). Otherwise, if too many iterations have occurred (perhaps 20), give up. Otherwise, go to step 2. (This is the end of the iteration loop.)

5. The covariance matrix of the parameters is

$$C_{gg} = N^{-1} - N^{-1}K^{\mathrm{T}}M^{-1}KN^{-1} \tag{5.38}$$

If this is the first time here, go to step 8.

6. For the last point tentatively rejected, recompute e_i and A_i as in step 2 using the latest values of the camera model parameters. Then,

$$r_i = e_i^{\mathrm{T}}(\sigma^2 I + A_i C_{gg} A_i^{\mathrm{T}})^{-1} e_i \qquad (5.39)$$

where I is the 2×2 identity matrix. If $r_i > 16$, reject this point. Otherwise, go to step 9 (exit from the edit loop).

7. If too many points have been rejected (perhaps 10) give up.

8. For each current point, use the most recent values in the following:

$$r_i = e_i^{\mathrm{T}}(\sigma^2 I - A_i C_{gg} A_i^{\mathrm{T}})^{-1} e_i . \qquad (5.40)$$

Tentatively reject the point with the greatest r_i. Then go to step 1. (This is the end of the edit loop.)

9. Reinstate the tentatively rejected point, and back up to the solution computed using this point. Use the results from this old solution below, and finish successfully.

If the run is successful, the camera model parameters g and their covariance matrix C_{gg} are the result.

A possible improvement that may be desirable in the case of grossly non-square pixels would be to compute separate q's for each image dimension by summing the squares of the discrepancies separately for x and y instead of using (5.33) and by dividing by $n-7$ instead of $2n-14$ in (5.37) to obtain the two variances. Then a 2×2 weight matrix with the reciprocal of these variances on the main diagonal would be included in the summations in (5.31) and (5.32) in the usual way [5], instead of factoring the variance out as above.

5.5   Use of Camera Model

5.5.1   Projecting from Object Space to Image Space

When a point p_i in three-dimensional space is available and it is desired to compute its projection into an image, p_i and the camera model computed in Sect. 5.4 are used in (5.3–5.12) in Sect. 5.2.2 to compute $\hat{x}_i$ and $\hat{y}_i$.

Often the partial derivatives of the above projection are needed (for error propagation or in a least-squares adjustment). These are obtained as follows:

$$\frac{\partial \hat{x}_i}{\partial p_i} = \frac{\partial \hat{x}_i}{\partial p_i'} \frac{\partial p_i'}{\partial p_i} \qquad (5.41)$$

$$\frac{\partial \hat{y}_i}{\partial p_i} = \frac{\partial \hat{y}_i}{\partial p_i'} \frac{\partial p_i'}{\partial p_i} \qquad (5.42)$$

in terms of the partial derivatives defined in Sect. 5.3. However, for most applications not much accuracy is needed in partial derivatives. Therefore, when the distortion is small ($|\rho_0| << 1, |\rho_1| << 1$, and $|\rho_2| << 1$) and speed is important, sufficient accuracy may be obtained by assuming that $\partial p_i'/\partial p_i$ is the identity matrix, so that the following results:

$$\frac{\partial \hat{x}_i}{\partial p_i} \approx \frac{\partial \hat{x}_i}{\partial p_i'} \tag{5.43}$$

$$\frac{\partial \hat{y}_i}{\partial p_i} \approx \frac{\partial \hat{y}_i}{\partial p_i'} . \tag{5.44}$$

Using these approximations can result in significant savings in time if many points are to be projected, since the computation of $\partial p_i'/\partial p_i$ is considerably more involved than is the computation of $\partial \hat{x}_i/\partial p_i'$ and $\partial \hat{y}_i/\partial p_i'$, as can be seen in Sect. 5.3.

5.5.2   Projecting from Image Space to Object Space

Sometimes a point in the image (x_i and y_i) is given and it is desired to project it as a ray in space (represented by the unit vector r_i). This can be done by first projecting the ray neglecting distortion, as follows: ray

$$r_i' = \text{sign}(a \cdot v \times h)\, \text{unit}[(v - y_i a) \times (h - x_i a)] \tag{5.45}$$

where sign $= \pm 1$ according to whether its argument is positive or negative. Then the distortion can be added by one of two methods.

In the first method, (5.3–5.7) are used with $p_i - c$ replaced by r_i and $p_i' - c$ replaced by r_i' (which is equivalent to considering a point at unit distance), to produce the following:

$$\zeta_i = r_i \cdot o \tag{5.46}$$

$$\lambda_i = r_i - \zeta_i \cdot o \tag{5.47}$$

$$\tau_i = \frac{\lambda_i \cdot \lambda_i}{\zeta_i^2} \tag{5.48}$$

$$\mu_i = \rho_0 + \rho_1 \tau_i + \rho_2 \tau_i^2 \tag{5.49}$$

$$r_i = \text{unit}(r_i' - \mu_i \lambda_i) \tag{5.50}$$

where we have had to rescale r_i to force it to be a unit vector, since otherwise the distortion correction changes only the component of the vector perpendicular to the optical axis, and thus changes the length of the vector. However, the equations in Sect. 5.2.2 were designed to go from the unprimed to the primed quantities, and here the opposite is desired. Therefore, in this method an iterative solution is done as follows: on the first iteration, r_i is set equal to r_i'; then an improved r_i is computed as above on each iteration. The values of the other quantities from the last iteration also are needed below if partial derivatives are computed.

The second method is faster and is the one that has been implemented. It can be derived by manipulating (5.46–5.50) to produce the following:

$$\zeta_i' = r_i' \cdot o \tag{5.51}$$

$$\lambda_i' = r_i' - \zeta_i' \cdot o \tag{5.52}$$

$$\tau_i' = \frac{\lambda_i' \cdot \lambda_i'}{\zeta_i'^2} \tag{5.53}$$

$$\rho_2 \tau_i'^2 (1 - \mu_i')^5 + \rho_1 \tau_i' (1 - \mu_i')^3 + (1 + \rho_0)(1 - \mu_i') - 1 = 0 . \tag{5.54}$$

Equation (5.54) can be solved for μ_i' by Newton's method (using $\mu_i' = 0$ as the initial approximation). Then,

$$r_i = \text{unit}(r_i' - \mu_i' \lambda_i') . \tag{5.55}$$

However, if partial derivatives are to be computed below, corrected values of $\zeta_i, \lambda_i, \tau_i$, and μ_i must be computed as in the first method, by using (5.46–5.49). (Now only one iteration is needed, since r_i has already been obtained.)

The partial derivatives of the projection into a ray can be obtained by differentiating (5.45) with respect to x_i and y_i to obtain the following:

$$\frac{\partial r_i'}{\partial x_i} = -\frac{\text{sign}(a \cdot v \times h)(I - r_i' r_i'^{\mathrm{T}})[(v - y_i a) \times a]}{|(v - y_i a) \times (h - x_i a)|} \tag{5.56}$$

$$\frac{\partial r_i'}{\partial y_i} = \frac{\text{sign}(a \cdot v \times h)(I - r_i' r_i'^{\mathrm{T}})[(h - x_i a) \times a]}{|(v - y_i a) \times (h - x_i a)|} . \tag{5.57}$$

Then the effect of distortion can be included by computing the partial derivatives as in Sect. 5.3, with $p_i - c$ replaced by r_i and with a rescaling because of the change in magnitude due to (5.50), to produce the following:

$$\frac{\partial \lambda_i}{\partial r_i} = I - oo^{\mathrm{T}} \tag{5.58}$$

$$\frac{\partial \mu_i}{\partial \tau_i} = \rho_1 + 2\rho_2 \tau_i \tag{5.59}$$

$$\frac{\partial r_i'}{\partial r_i} = \left[I + \frac{\partial \mu_i}{\partial \tau_i} \lambda_i \left(\frac{2}{\zeta_i^2} \lambda_i^{\mathrm{T}} \frac{\partial \lambda_i}{\partial r_i} - \frac{2\tau_i}{\zeta_i} o^{\mathrm{T}} \right) + \mu_i \frac{\partial \lambda_i}{\partial r_i} \right] |r_i' - \mu_i \lambda_i| . \tag{5.60}$$

Then, since the transformation in the other direction is desired, the inverse of the matrix is used, as follows:

$$\frac{\partial r_i}{\partial x_i} = \left(\frac{\partial r_i'}{\partial r_i} \right)^{-1} \frac{\partial r_i'}{\partial x_i} \tag{5.61}$$

$$\frac{\partial r_i}{\partial y_i} = \left(\frac{\partial r_i'}{\partial r_i} \right)^{-1} \frac{\partial r_i'}{\partial y_i} . \tag{5.62}$$

As in projecting in the other direction, this correction of the partial derivatives for distortion can be omitted when speed is important and not much accuracy is needed, if the distortion is small.

136   Donald B. Gennery

Acknowledgements

The research described herein was carried out by the Jet Propulsion Laboratory, California Institute of Technology, under contract with the National Aeronautics and Space Administration. The camera calibration program was written by Todd Litwin.

References

1. Y. Yakimovsky, R. T. Cunningham. A System for Extracting Three-Dimensional Measurements from a Stereo Pair of TV Cameras. Computer Graphics and Image Processing **7**, 195–210, 1978
2. D. B. Gennery, T. Litwin, B. Wilcox, B. Bon. Sensing and Perception Research for Space Telerobotics at JPL. *Proc. IEEE International Conference on Robotics and Automation*, Raleigh, NC, March 31–April 3, pp. 311–317, 1987
3. D.B. Gennery. Camera Calibration Including Lens Distortion, JPL internal report D-8580, Jet Propulsion Laboratory, Pasadena, CA, 1991
4. F.A. Jenkins, H.E. White. *Fundamentals of Optics*, McGraw-Hill, 1950
5. E.M. Mikhail (with contributions by F. Ackermann). *Observations and Least Squares*, Harper and Row, 1976
6. D.B. Gennery. Modelling the Environment of an Exploring Vehicle by Means of Stereo Vision, AIM-339 (STAN-CS-80-805), Stanford University, Computer Science Dept., June 1980.
7. D.B. Gennery. Stereo Vision for the Acquisition and Tracking of Moving Three-Dimensional Objects. In *Techniques for 3-D Machine Perception*, A. Rosenfeld (ed.) North-Holland, Amsterdam, 1986

6 Modeling and Calibration of Variable-Parameter Camera Systems

Reg G. Willson and Steven A. Shafer

Summary

Camera systems with adjustable lenses are inherently more useful than those with fixed lenses. Adjustable lenses enable us to produce better images by matching the camera's sensing characteristics to the conditions in a scene. They also allow us to make measurements by noting how the scene's image changes as the lens settings are varied. The reason adjustable lenses are not more commonly used in machine vision is that they are difficult to model.

In this chapter we demonstrate how two commonly used camera models – the pinhole camera model and the – cannot be easily extended to cameras with adjustable lenses. We then show how an approach can provide us with accurate models for such systems. We conclude by presenting the problem of camera calibration in the context of a more general top-down philosophy of modeling and calibration.

6.1 Abstract Camera Models

In machine vision we need to know aspects of a camera's image-formation process that range from simple properties, such as magnification and focussed distance, to more complex image properties, such as perspective projection and image defocus. In order to have computationally efficient, closed-form equations for the more complex properties we use models that are based on simplifications or abstractions of the lens's true image-formation process. The two most commonly used abstract models are the pinhole camera model and the thin-lens model.

In the basic pinhole camera model, illustrated in Fig. 6.1, the 3D coordinates of a point $P(x_c, y_c, z_c)$ are related to the 2D coordinates of its image $P'(X_u, Y_u)$ by the equations

$$X_u = f\frac{x_c}{z_c},$$
$$Y_u = f\frac{y_c}{z_c}$$

where f is the pinhole camera model's effective focal length.

In the thin-lens model, illustrated in Fig. 6.2, the position of a point P in front of the lens is related to the position of the point's focussed image P'

Springer Series in Information Sciences, Vol. 34
Calibration and Orientation of Cameras in Computer Vision
Eds.: Gruen, Huang © Springer-Verlag Berlin Heidelberg 2001

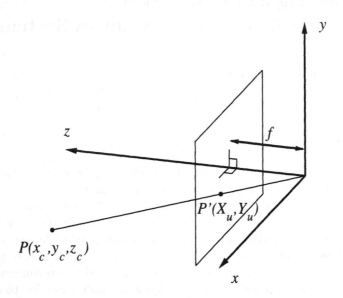

Fig. 6.1. Basic pinhole image-formation model.

behind the lens by the equation

$$\frac{1}{s} + \frac{1}{s'} = \frac{1}{f}$$

where f is the thin-lens model's focal length, s is the object to lens separation, and s' is the lens to focussed image plane separation.

For cameras with fixed lenses the model parameters (e.g. f, s, and s') are constants. To calibrate the camera model all we need to do is find values for the constants.

For cameras with adjustable lenses the model parameters vary with the lens's actuators. To calibrate the camera model we need to formulate and calibrate the functions that describe the relationships between the model parameters and the actuator settings. As we shall see, these relationships are not as simple or direct as the abstract camera models would seem to suggest.

6.1.1 Building Multi-Degree-of-Freedom Camera Models

In variable-parameter lenses there is generally, by design, a nearly 1:1 correspondence between the lens's control parameters (e.g. focus or zoom) and a specific property of the lens's image-formation process (e.g. focussed distance or magnification). This correspondence is often the basis for models for systems where only one continuous control parameter is to be used. Unfortunately, when building models for systems having two or more degrees-of-freedom (DOFs) these first-order relationships provide little insight into the interactions between the control parameters and the lens's image-formation process.

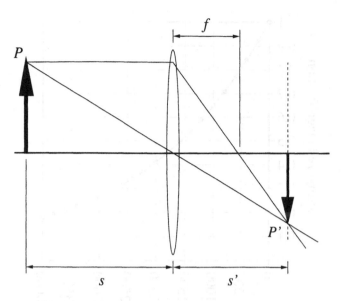

Fig. 6.2. Basic thin-lens image-formation model.

In this section we develop multi-degree-of-freedom (MDOF) models for three simple image formation properties: image magnification, focussed distance, and image center. For each model we start by describing a 1-DOF model based on the first-order lens behavior. We then continue by augmenting the model with additional control parameters whenever experimental evidence indicates they are significant.

Building a MDOF Magnification Model. Image magnification can be defined as the ratio of image size to object size. Using the pinhole camera model of the image formation process our magnification model can be expressed as

$$M = cf$$

where c is a constant of proportionality and f is the lens's effective focal length.

For variable-parameter camera systems the lens's focal length is controlled by the zoom motor. Plotting the normalized image magnification as a function of the zoom motor (Fig. 6.3) we see that for our lens there is a reasonably direct relationship between the motor values and our model's effective focal length parameter. If we model this relationship with a simple second-order polynomial our 1-DOF magnification model can be expressed as

$$M = c_1 + c_2 m_z + c_3 m_z^2$$
$$= g_M(m_z; c_1, c_2, c_3)$$

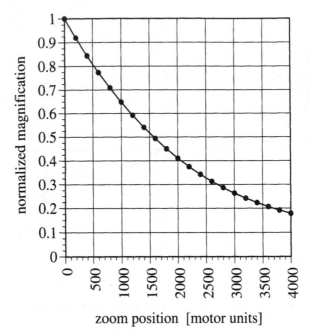

Fig. 6.3. Magnification vs zoom motor.

where M is the image magnification, g_M is the parameterized model of the control parameter and the calibration coefficients, m_z is the zoom motor, and c_1, c_2, c_3 are coefficients determined by calibration.

Unfortunately our 1-DOF model doesn't tell the whole story. For most lenses image magnification can also be changed by adjusting the lens's focus. Figure 6.4 shows a plot of image magnification as a function of the lens's focus motor for our lens. The reason for the change in magnification can be inferred from the thin-lens model of the image formation process, illustrated in Fig. 6.5. With this model as the focus of the lens is varied from near to far the magnification of an object located at a fixed distance from the lens decreases.

As the pinhole camera model has no concept of focus there does not seem to be any direct way to augment our existing 1-DOF model with the focus DOF. However, if we again consider the thin-lens model of the image formation process illustrated in Fig. 6.2 we see that there should be a direct relationship between the lens's focus position, s', and the image magnification. Thus to augment the 1-DOF model possibly all we need do is multiply it by the lens's focus position. If we model the relationship between the lens's focus position and the focus motor with another second-order polynomial the resulting 2-DOF magnification model can be expressed as

$$M = (c_1 + c_2 m_z + c_3 m_z^2)(c_4 + c_5 m_f + c_6 m_f^2)$$
$$= g_M(m_z, m_f; c_1, ..., c_6)$$

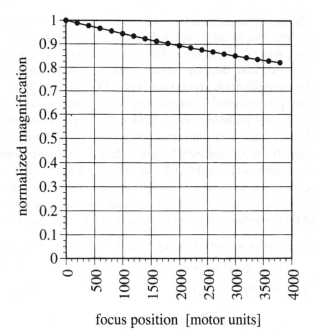

focus position [motor units]

Fig. 6.4. Magnification vs focus motor.

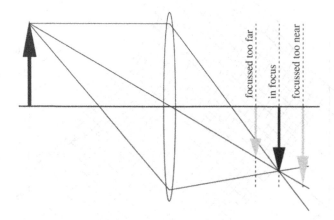

Fig. 6.5. Focus magnification.

where M is the image magnification, g_M is the parameterized model of the control parameters and calibration coefficients, m_z is the zoom motor, m_f is the focus motor, and $c_1, \ldots, c_6$ are coefficients determined by calibration.

At this point we find that the above form for our magnification model doesn't fit the magnification data that well. This occurs for two reasons. First, the actual process involved in varying a zoom lens's focal length really has no relation to the pinhole camera model. The second reason is that the relationship between the lens's motor settings and its internal hardware configuration is essentially an arbitrary function chosen by the lens manufacturer and is not necessarily well modeled by simple quadratic functions. To deal with these problems a more generalized modeling technique must be used. In [1] we use bivariate, cubic polynomials to fit the magnification data. The resulting 2-DOF magnification model, plotted in Fig. 6.6, has the form

$$M = c_1 + c_2 m_f + c_3 m_f^2 + c_4 m_f^3 + c_5 m_z + c_6 m_z m_f + c_7 m_z m_f^2$$
$$+ c_8 m_z^2 + c_9 m_z^2 m_f + c_{10} m_z^3$$
$$= g_M(m_z, m_f; c_1, \ldots, c_{10})$$

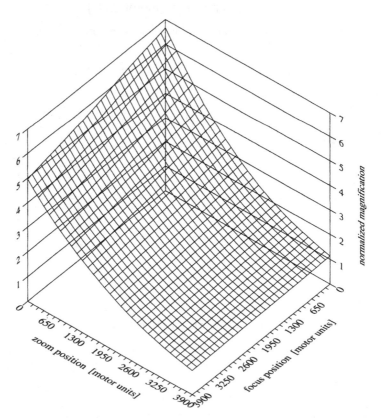

Fig. 6.6. Magnification vs focus and zoom motors.

where M is the image magnification, g_M is the parameterized model of the control parameters and calibration coefficients, m_z is the zoom motor, m_f is the focus motor, and $c_1, \ldots, c_{10}$ are coefficients determined by calibration.

Our model is not complete, however. If our lens behaved ideally the aperture would have no effect on image magnification. In practice the lens's aperture diameter influences the lens's primary aberrations, which in turn can affect the lens's magnification. Figure 6.7 shows a 0.26% drop in the image magnification as a function of lens aperture for one particular focus and zoom setting. From the graph of the relationship it would seem reasonable to model this variation with yet another cubic polynomial, expanding our magnification model into a trivariate, cubic polynomial. The resulting 3-DOF magnification model now has the form

$$M = g_M(m_z, m_f, m_a; c_1, \ldots, c_{19})$$

where M is the image magnification, g_M is the parameterized model of the control parameters and calibration coefficients, is the zoom motor, m_f is the focus motor, m_a is the aperture motor, and $c_1, \ldots, c_{19}$ are coefficients determined by calibration.

Our model is *still* not complete, however. Ideally, the lens's image magnification would be independent of the wavelength of the light used for the imaging. This is not the case in reality. Lateral chromatic aberration, illustrated in Fig. 6.8, can cause the magnification of images taken in different color bands to differ by as much as 0.5% for our lens [1]. Figure 6.9 shows the image magnification for the red, green, and blue bands for several focus motor positions. It would be virtually impossible to find a continuous closed form for the dependence of the magnification on wavelength for a lens. Fortunately, most multi-band imaging systems use only a small, fixed number of color filters to select the image band so we can add an image band DOF to the model by simply calibrating a separate magnification model in each band with data taken in that band. Our final 4-DOF magnification model now looks like

$$M = g_M(m_z, m_f, m_a, m_b; c_1, \ldots, c_{57})$$

where M is the image magnification, g_M is the parameterized model of the control parameters and calibration coefficients, m_z is the zoom motor, m_f is the focus motor, m_a is the aperture motor, m_b is the image band, and $c_1, \ldots, c_{57}$ are three sets of coefficients determined by calibration.

Building a MDOF Focussed Distance Model. The focussed distance of a lens can be defined as the distance between the plane of best focus in object space and some reference point in the camera system. In our work we use the location of the camera's sensor plane as the reference point because its position is physically measurable and because in most camera systems its position remains fixed relative to the camera body as the camera parameters

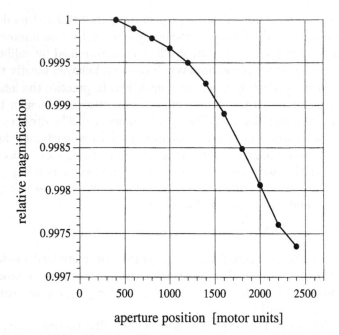

Fig. 6.7. Magnification vs aperture motor.

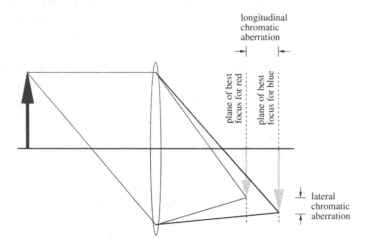

Fig. 6.8. Chromatic aberration.

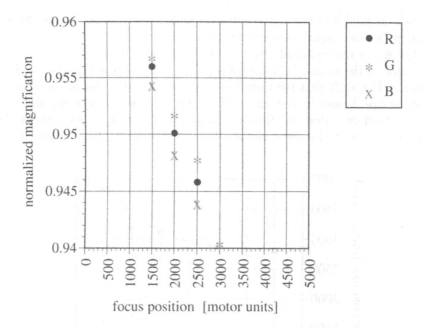

Fig. 6.9. Magnification vs focus motor and image band.

are varied. In all imaging systems the sensing plane has a fixed finite spatial resolution which gives rise to a zone called the depth of field in which all objects appear to be in focus. There is no distinct plane of best focus. To deal with this we define the focussed distance of a lens to be the distance between a target and the camera's sensor plane at which the target's sharpness (as measured by some sharpness criterion) peaks. Many sharpness criterion functions have been proposed for machine vision systems. For our work we use the common sum of squared image gradients.

Using the thin-lens model of the image formation process the lens's focussed distance (with the sensor as the reference point) can be expressed as

$$L = \frac{(s')^2}{s' - f}$$

where s' is the distance between the lens and the sensor plane and f is the lens's focal length. So, for our 1-DOF focussed distance model all we need do is supply a model for the relationship between the focus motor and the s' term.

Looking more closely at the thin-lens model we see that the lens's focussed distance should be a function of the lens's zoom setting as well as the focus setting. But modern zoom lenses are typically designed to be equifocal. That is, the focussed distance of the lens is designed to remain constant as the lens's focal length or zoom control is varied. This represents a fundamental inconsistency between our first-order lens model and the real lens behavior.

Without a good idea of what is going on within the lens we are now forced to adopt a more empirical modeling approach.

In spite of the equifocal design of the zoom lens, if we take careful measurements of the focussed distance of the lens as the zoom is varied we find, as shown in Fig. 6.10, that the focus motor position for the sharpness criterion function's peak does indeed vary with changes in the lens's zoom, although not to the degree that our thin-lens model would suggest. As a result the zoom parameter will have to be included in the focussed distance model.

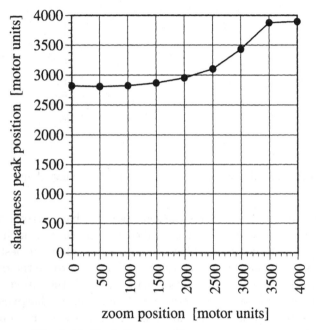

Fig. 6.10. Peak focus position vs zoom motor.

In an ideal lens the position of the plane of best focus should be unaffected by changes in the lens's aperture. Again though, if we take measurements we find that the position of sharpest focus for the lens is also a function of the lens aperture. Figure 6.11 shows the variation in the sharpest focus motor position with changes in the lens aperture. Part of this variation may be due to the effects of asymmetric changes in the depth of field with aperture, and part may be due to spherical aberration within the lens. In either event the aperture parameter will need to be included in the model.

Checking out the final DOF, the spectral band, we find that like the magnification model the lens's focussed distance is in fact influenced by the wavelength of the light used in imaging. This difference in focussed distance is the result of longitudinal chromatic aberration, illustrated in Fig. 6.8. So, like our MDOF magnification model, the focussed distance model must also

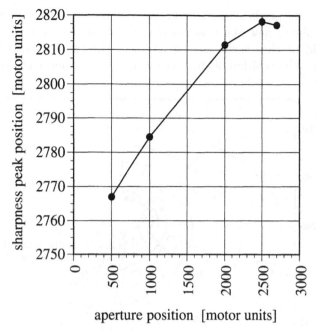

Fig. 6.11. Peak focus position vs aperture motor.

be parameterized with the image band. In the end our final 4-DOF focussed distance model has the form

$$L = g_L(m_z, m_f, m_a, m_b; c_1, \dots, c_n)$$

where L is the focussed distance, g_L is the parameterized model of the control parameters and calibration constants, m_z is the zoom motor, m_f is the focus motor, m_a is the aperture motor, m_b is the image band, and $c_1, \dots, c_n$ are coefficients determined by calibration.

Building a MDOF Image Center Model. In an ideal lens the image center is considered to be the point of intersection of the lens's optical axis with the camera's sensing plane. In fact there are many possible definitions of image center, and in real lenses most do not have the same coordinates. Image centers also move as lens parameters are changed. While the definition used for image center and the accuracy of its measurement can be an important factor in the accuracy of the overall camera calibration, for many machine vision applications the principal concern is the motion of the image center as the lens parameters are varied. For our applications we use a model that describes the relative shift in the camera's image center as a function of the lens control parameters.

In both the pinhole and thin-lens camera models the image center is considered to be a fixed parameter. However, in real lenses the image center may shift several pixels as the lens's focus and zoom parameters are varied. Figure 6.12 shows how the position of a fixed point at the center of our camera's field of view varies as a function of the camera lens's focus and zoom motors. The position of the image center also varies with image band.

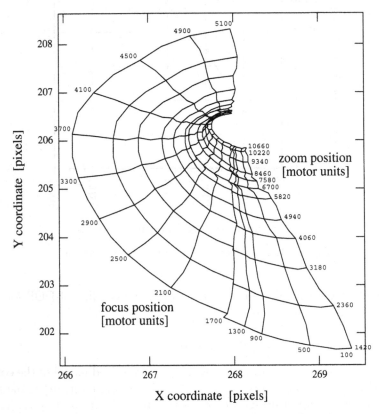

Fig. 6.12. Image center as a function of lens focus and zoom motors.

The explanation for the variation in image center is the unavoidable misalignment of the optical components on the lens that results from manufacturing tolerances. Changing the positions of lens components relative to the sensing plane, either by adjusting the focus or the zoom, causes changes in the misalignment which in turn causes the image center to move. In addition, the slightly different optical paths through the lens taken by light of different wavelengths means that for any given lens position each image band will not necessarily have the same image center. Since a theoretically derived model for the misalignment in the optical components would be intractable we quickly go to our empirical approach, basing the model on measurements taken directly from the system. The result is a 3-DOF model of the form

$$(x_0, y_0) = g_{x_0 y_0}(m_z, m_f, m_b; c_1, \ldots, c_n)$$

where (x_0, y_0) is the location of the image center, $g_{x_0 y_0}$ is the parameterized model of the control parameters and calibration constants, m_z is the zoom motor, m_f is the focus motor, m_b is the image band, and $c_1, \ldots, c_n$ are coefficients determined by calibration.

The only remaining DOF in the camera system that can potentially have an effect on the image center is the aperture. For our lenses, careful measurements show no significant shift in the image center with changes in the aperture. If some dependency does exist, it's below our measurement capability. So, we accept the above MDOF model of image center as complete.

In [1] we have noted where the results for two vision tasks traditionally done using simple first-order or 1-DOF models of lens behavior, color imaging and focus ranging, can be compromised by unmodelled lens behaviors. By developing more detailed MDOF models we have been able to find a way to control the lens to achieve better image quality and better results in the vision tasks. We now present those problems and their solution.

6.1.2   Applying MDOF Camera Models to Vision Tasks

Many vision tasks make use of sequences of images taken while one of the imaging system control parameters is varied. For example, in color imaging typically three images of the same scene are taken in three different color bands. For static scenes this can be done by changing a color filter in front of the camera. In focus ranging sequences of images are taken at different focussed distances by varying the lens's focus control. In both these tasks there is an underlying assumption about the relationship between the changes in the camera's controls and the changes in the resulting image parameters. For the most part these assumptions are based on first-order approximations of the camera lens's behavior. In this section we illustrate how the assumption of first-order lens behavior causes problems for both the color-imaging and focus-ranging tasks. Subsequently we show how we can significantly improve the performance of both tasks by using more accurate MDOF models of lens behavior.

MDOF Color Imaging. Color-image analysis uses the information contained in three spectral bands to determine properties of the scene being imaged. Implicit in any color-image analysis is the assumption that the per-pixel information in each band corresponds to the same point, region, or volume in object space. As we will demonstrate in the following section, this is not always so.

To simplify color band alignment, virtually all color imaging processes use a single lens with color filters being used to separate out the different bands. In this situation the color band (i.e. filter) can be considered to be a single DOF which is varied to accomplish the imaging task. To a first order approximation we expect that since the focus, zoom, and aperture of the camera are unchanged while the images are taken, the magnification, focussed distance, and centering of the three images will also be unchanged. As we have described in the previous section though, this first-order approximation of the lens behavior is not necessarily that good. In this case the second-order effect of chromatic aberration may cause significant magnification, focus, and centering differences between each of the color bands.

Instead of the 1-DOF color imaging approach, if we use the MDOF camera models g_M, g_L and $g_{x_0y_0}$ we can determine the control parameter values required to take images that have the same magnification, focussed distance and image center. Given the control parameter values for the blue image $(m_{z_b}, m_{f_b}, m_{a_b}, m_{b_b})$, we can find the correct control parameter values for the red and green images by satisfying the image parameter constraints

$$g_M(m_{z_r}, m_{f_r}, m_{a_b}, m_{b_r}) = g_M(m_{z_g}, m_{f_g}, m_{a_b}, m_{b_g}) = g_M(m_{z_b}, m_{f_b}, m_{a_b}, m_{b_b})$$

and

$$g_L(m_{z_r}, m_{f_r}, m_{a_b}, m_{b_r}) = g_L(m_{z_g}, m_{f_g}, m_{a_b}, m_{b_g}) = g_L(m_{z_b}, m_{f_b}, m_{a_b}, m_{b_b}).$$

To correct for decentering between the bands we determine the difference in the image centers in the camera's image space

$$\text{shift}_{rb} = g_{x_0y_0}(m_{z_r}, m_{f_r}, m_{a_b}, m_{b_r}) - g_{x_0y_0}(m_{z_b}, m_{f_b}, m_{a_b}, m_{b_b})$$
$$\text{shift}_{gb} = g_{x_0y_0}(m_{z_g}, m_{f_g}, m_{a_b}, m_{b_g}) - g_{x_0y_0}(m_{z_b}, m_{f_b}, m_{a_b}, m_{b_b})$$

and then shift the camera in object space, parallel to the sensor plane, to compensate.

Table 6.1 lists a typical set of control parameter values for a corrected set of color images taken using one of our camera systems.

Table 6.1. Typical camera and lens settings for corrected color images.

Color filter	Focus motor (out of 3900)	Difference from blue setting	Zoom motor (out of 4000)	Difference from blue setting	Camera X shift
Red	1407	−6.36%	533	+0.83%	0.15 mm
Green	1592	−1.62%	508	+0.20%	0.05 mm
Blue	1655		500		

To determine the effectiveness of the MDOF color imaging approach we measure the misregistration between the image bands of a color picture taken of a black on white checkerboard test target. The basic procedure involves accurately locating the positions of the checkerboard's black to white step edges to subpixel accuracy in each of the bands. The relative positions of the same black to white edges in each band then provide a measure of the chromatic aberration induced misregistration at that point in the image. By using a relatively dense set of vertical and horizontal black to white edges (e.g. the checkerboard pattern) we can approximate the misregistration at any point in the image by using the nearest pairs of horizontal and vertical step edges to determine the orthogonal components of the misregistration.

Figures 6.13 and 6.14 show the magnitude of the image misregistration (in pixel widths) between the blue and red images coded in a gray scale ranging from 0 pixels (black) to 1.4 pixels (white). Figure 6.13 shows the blue-red misregistration for the target imaged without lens compensation. The misregistration across the image ranges from 0 to 1.2 pixel widths. For the blue-red case the zero error region is slightly to the right of the image center. In general the location of the zero error region for the blue-green case will not be the same. Figure 6.14 shows the magnitude of the blue-red misregistration for the target imaged with lens compensation. The remaining misregistration is now less than 0.1 pixel widths over most of the image. This result indicates that with active lens compensation it is possible to reduce the image band misregistration introduced by chromatic aberration by over an order of magnitude.

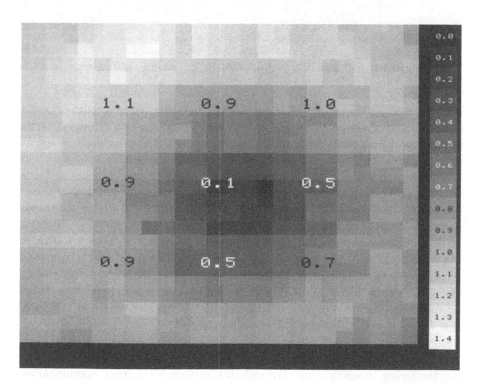

Fig. 6.13. Full field blue/red misregistration – uncompensated image.

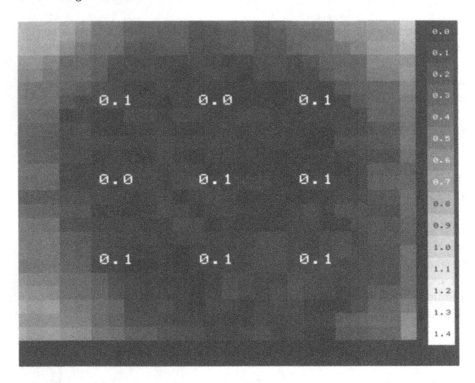

Fig. 6.14. Full field blue/red misregistration – compensated image.

MDOF Focus Ranging. In focus ranging the distance to a feature in the camera's field of view is estimated by moving the camera lens through a series of focus positions while determining the sharpness of the feature's image at each position. The feature's sharpness is measured using a criterion function evaluated over a region of the image containing the feature. Given the focus position of the lens where the criterion function peaks, the distance between the camera and the feature in the image can be estimated using a calibrated focussed distance model.

In conventional focus ranging the sequence of images is taken while varying a single DOF – the lens's focus. As we have described in the previous section, varying the focus of the camera's lens causes the second-order effect of focus magnification. As described by Willson [1], these changes in image magnification can lead to a target dependent bias in the position of the criterion function peak. This bias in the position of the criterion function peak can result in range estimate errors in the order of several percent. To eliminate this bias we need to vary the focussed distance of the lens without changing the image magnification. We call this approach constant magnification focussing. By using the MDOF camera models g_M and g_L we can determine the control parameter values required to take the constant magnification images. Given a starting set of control parameter values m_{z_s}, m_{f_s}, m_{a_s}, and

m_{b_s}, we can find a sequence of control parameter values for the remaining images by satisfying the image parameter constraint

$$g_M(m_{z_s}, m_{f_s}, m_{a_b}, m_{b_s}) = g_M(m_{z_1}, m_{f_1}, m_{a_b}, m_{b_s})$$
$$= , \ldots, . = g_M(m_{z_n}, m_{f_n}, m_{a_b}, m_{b_s}),$$

where the values of m_{f_i} are chosen by some focus search strategy.

Having found the lens control parameters at which the criterion function peaks we simply use the focussed distance model g_L to determine the feature's range. Note that we actually require two MDOF camera models, g_M and g_L, for the MDOF focus ranging approach. The first MDOF model is required to implement constant magnification focussing. The second MDOF model is required because constant magnification focussing finds the criterion function peak while varying both the focus and zoom.

To illustrate the effects of focus magnification on the position of our criterion function peak we use two targets, illustrated in Fig. 6.15. Target 1 is a black square or box on a white background, completely enclosed in the evaluation window. Target 2 is a black bar on a white background with only the center region of the bar contained in the evaluation window. For both targets the majority of the value of the criterion function results from the black to white edge with the totally white and totally black regions contributing insignificant amounts. For the box target the length of the perimeter of the black region changes with the focus magnification, while for the bar target the perimeter length remains constant. The box and bar targets are both located on the same plane 1.5 m away from the camera's sensor plane.

For the conventional 1-DOF focussing the focussed distance motor is varied while the zoom motor is held constant at 500 motor units. The image magnification over the range of focus changes by a factor of 0.751. For MDOF constant magnification focussing the focussed distance motor is varied while the zoom motor is concurrently varied to keep the magnification constant. Figure 6.16 shows the zoom motor settings versus the focus motor settings for both the compensated (∗) and the uncompensated (●) focussing. For compensated focussing the zoom motor settings are varied from 747 motor units to 91 motor units. The constant magnification focussing curve is essentially an isomagnification contour from the magnification model shown in Fig. 6.6.

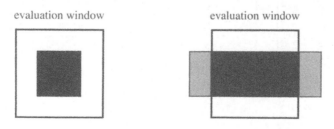

Fig. 6.15. Box and bar focus targets.

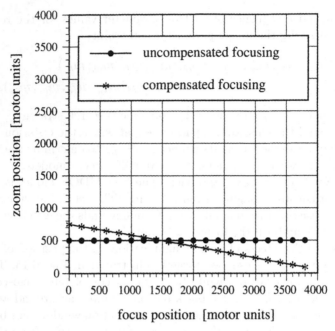

Fig. 6.16. Focus vs zoom settings for both 1-DOF (●) and MDOF (*) focussing.

Figure 6.17 shows that with conventional focussing there is a significant bias in the peak position for the box. When the focus magnification is compensated for the bias in the peak position for the box target is eliminated.

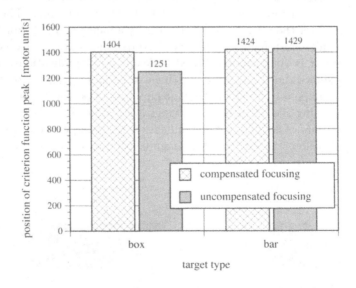

Fig. 6.17. Criterion function peak position vs target type and focus type.

The peak position for the bar target is unaffected. For an actual range of 1.5 m the bias in the uncompensated box target's peak position corresponds to a 6% error in the range measurement determined from the calibrated lens model.

6.1.3   A General Theory of Camera Modeling and Calibration

In the preceding sections, we presented models and calibration data for various aspects of imaging with a MDOF camera system. As we showed, the traditional approach of using parameterized, closed-form equations based on the underlying physics is inadequate for capturing the behavior of such systems. Instead, we used more general polynomials that fit the observed data, without relying on the specific forms of the equations that arise from the standard optics. However, these models were presented a piece at a time, without an integrating framework. Historically, this is how our research developed. Naturally, we were led to speculate whether there could be a more systematic and comprehensive formulation of the modeling and calibration task that would encompass all of these individual elements. We now present such a general framework and discuss its implications for MDOF camera modeling and calibration.

We begin by noting that the problem introduced by reliance on the idealized thin-lens model is that it predicts a perfect behavior, while a real MDOF camera exhibits many imperfections and aberrations, most of which are arbitrary in nature. Therefore, in the theoretical limit, every characteristic of the image might depend in an arbitrary way on every controllable parameter of the camera system. This is a useful starting point for the general theory of modeling and calibration. We can characterize the image data as the discrete array $P(r, c)$, where P is measured pixel value, and (r, c) are the row and column coordinates of each pixel. This is the output of the imaging process. The input is the set of rays of light $L(X, Y, \Theta, \Phi; \lambda)$ entering the lens, where (X, Y) are world coordinates on the front nodal plane $(Z = 0)$ of the lens; Θ and Φ indicate the angle of the incoming ray in spherical coordinates; and λ is the wavelength of light. L itself represents the energy (radiance) of the ray; it is a well-known function that has been called the "plenoptic function", the "helios", the "photonic field", and other names. The functions L and P represent the input and output of the imaging system.

The Generalized Responsivity and Measurement Functions. The goal of modeling and calibration is to find the mapping from L to P. In the most general case, it could be an arbitrary functional on L. But, we can be fairly safe in assuming, for most electronic camera systems, that the mapping begins with the integration of light energy on a sensor plane, followed by the (possibly nonlinear) electronic measurement of the integrated energy. This can be expressed mathematically by defining a quantity $I(r, c)$ which represents the integrated energy contributing to the pixel $P(r,c)$. Then we can express I as a weighted integral of L:

$$I(r,c) = \int R(r,c;X,Y,\Theta,\Phi;\lambda)\, L(X,Y,\Theta,\Phi;\lambda)\, \mathrm{d}X\, \mathrm{d}Y\, \mathrm{d}\Theta\, \mathrm{d}\Phi\, \mathrm{d}\lambda\,.$$

Similarly, the pixel value P results from some measurement of I:

$$P(r,c) = M(r,c;I)\,.$$

In these equations, R represents the responsivity of the camera in terms of both geometric and radiometric parameters; M represents the process whereby the integrated energy is converted to a pixel value. Note that this is a very general formulation; for example, it does not assume perfect focussing or perspective projection as in the thin-lens model.

The equations above model the imaging process very generally, but they do not address the effects of varying the parameters of a MDOF system. Let us for the moment assume the parameters we can control are the lens parameters m_z, m_f, and m_a; and the color filter band denoted by m_b. These form a four-dimensional "control space" for the camera system. The functions R and M may be different at each point in that space; thus, the control parameters modulate R and M, and may be inserted as additional parameters:

$$I(r,c)$$
$$= \int R(r,c;m_z,m_f,m_a,m_b;X,Y,\Theta,\Phi;\lambda)\, L(X,Y,\Theta,\Phi;\lambda)\, \mathrm{d}X\, \mathrm{d}Y\, \mathrm{d}\Theta\, \mathrm{d}\Phi\, \mathrm{d}\lambda$$

$$P(r,c) = M(r,c;m_z,m_f,m_a,m_b;I)$$

We call this function R the "generalized responsivity function" and this M the for the system.

The goal of camera modeling and calibration is then to determine the form and constants in these functions R and M.

Reduction to Common Forms. The commonly used equations of imaging are very simple: (X,Y,Z) is projected onto the image point (x,y) by

$$(x,y) = \left(\frac{fX}{f-Z}, \frac{fY}{f-Z} \right)$$

$$P(r,c) = a^2 \int L(x,y,\lambda)\, S(\lambda)\, \tau_b(\lambda)\, \mathrm{d}x\, \mathrm{d}y\, \mathrm{d}\lambda$$

with these assumptions: (1) Z is the focussed distance of the lens, determined by m_f; (2) f is the focal length, controlled by m_z; (3) a is the aperture, controlled by m_a; (4) the integration limits for x and y in the second equation are determined by the bounds of the pixel location (r,c); (5) $S(\lambda)$ is the sensor's spectral responsivity; (6) $\tau_b(\lambda)$ is the transmittance of color filter m_b.

How are these simple imaging equations related to the functions R and M presented above? It can be shown (by a sometimes laborious development not presented here) that, if many idealizing assumptions are made, then R and M can be reduced to yield exactly these simple equations. The reduction process can be summarized as follows:

- Spectral reduction of R: Assume there is a color filter selected by control parameter m_b, with spectral transmittance $\tau_b(X, Y, \Theta, \Phi; \lambda)$. The variables X, Y, Θ, Φ express the possibility that the spectral transmittance varies somewhat with the position and direction of each ray passing through the filter. If we postulate that such variations do not exist, we can simply write $\tau_b(\lambda)$. Similarly, for the sensor, we might write $S(X, Y, \Theta, \Phi; \lambda)$ and simplify it to $S(\lambda)$. If we assume that these represent the only spectral variation in the camera response, then we can remove the variables m_b and λ from R and rewrite the first equation as a product of geometric and spectral terms:

$$I(r, c) = \int R(r, c; m_z, m_f, m_a; X, Y, \Theta, \Phi)$$

$$L(X, Y, \Theta, \Phi; \lambda)\, S(\lambda)\, \tau_b(\lambda)\, \mathrm{d}X\, \mathrm{d}Y\, \mathrm{d}\Theta\, \mathrm{d}\Phi\, \mathrm{d}\lambda$$

The simplification may not be obvious at first glance, but note that the number of parameters of R has been reduced from eleven to nine.

- Geometric reduction of R: This involves three coordinate systems: (X,Y,Z) in the world (with $Z = 0$ on the front nodal plane of the lens); (X', Y', Z') at the rear nodal plane of the lens; and (x, y, z) on the sensor plane (with $z = 0$ on the sensor plane itself). The transformation from X-Y-Θ-Φ to X'-Y'-Θ'-Φ' is based on the focal length and thin-lens model (note that the hiatus in the thick-lens model is simply the distance along Z between the two nodal planes, thus is irrelevant once X'-Y'-Z' coordinates are achieved). A ray X'-Y'-Θ'-Φ' is transformed to x-y-θ-ϕ based on the focussed distance (i.e. lens-to-sensor distance s'); ignoring θ and ϕ, we have the x-y location at which the original incoming ray X-Y-Θ-Φ strikes the sensor plane. Therefore, we can substitute variables in the above equation, yielding:

$$I(r, c) = \int R(r, c; m_z, m_f, m_a; x, y)\, L(x, y; \lambda)\, S(\lambda)\, \tau_b(\lambda)\, \mathrm{d}x\, \mathrm{d}y\, \mathrm{d}\lambda \ .$$

Note that the equations relating X-Y-Θ-Φ to x-y subsume the usual perspective projection. If we further postulate that this is the only effect of the zoom and focus, we can eliminate those variables from R; and if we assume the source points X-Y-Z are uniform point-source emitters, we do not need to explicitly consider R to vary over x and y. This yields:

$$I(r,c) = R(r,c;m_a) \int L(x,y;\lambda)\, S(\lambda)\, \tau_b(\lambda) \mathrm{d}x\, \mathrm{d}y\, \mathrm{d}\lambda$$

- Radiometric reduction of R: Now, we can assume that the only effect of the aperture a (controlled by m_a) is that the aperture area (proportional to a^2) determines a scale factor which is the same for all pixels in the image. Then, we have:

$$I(r,c) = a^2 \int L(x,y;\lambda)\, S(\lambda)\, \tau_b(\lambda)\, \mathrm{d}x\, \mathrm{d}y\, \mathrm{d}\lambda$$

- Reduction of M: We can be safe in assuming that the lens and filter parameters do not affect the electronic process of converting integrated energy to the final pixel value, $P(r,c) = M(r,c;\, I)$. If we assume M is uniform for all pixels, we can further simplify to: $P(r,c) = M(I(r,c))$. It is commonly assumed that $M(I) = I^\gamma$ for some γ; if we make this assumption and further assume linearity ($\gamma = 1.0$), then $P(r,c) = I(r,c) = a^2 \int L(x,y,\lambda)\, S(\lambda)\, \tau_b(\lambda)\, \mathrm{d}x\, \mathrm{d}y\, \mathrm{d}\lambda$, and we have at last the simple imaging equation commonly assumed.

In this derivation, we have assumed that each incoming ray X-Y-Θ-Φ is mapped onto a unique ray at the sensor plane; this is fairly "safe" for real systems. We have ignored effects such as variation over time, polarization, and controllable exposure time and/or gain in the camera system. All of these effects can be incorporated into the model above to yield a more comprehensive, though more complex, formulation of R and M. However, these additional parameters can be postulated to simplify and therefore drop out of the functions, resulting in the same simple models presented above.

The Modeling and Calibration Process. Because R and M are (if properly formulated) completely general, they must capture the true behavior of the imaging system. They can be reduced by a chain of many assumptions to the simple, commonly used, idealized equations of imaging. As we have seen in this chapter, those equations are generally inaccurate for a real MDOF system. The problem lies in the long chain of assumptions that bridge from the general form to the idealized model. Some of these assumptions are fairly likely to be true; for example, the gamma factor γ and the lens focal length f are determined by very different parts of the imaging system, and thus these two are assuredly independent.

On the other hand, some assumptions are equally sure to be false in a real system. For example, we assumed that the spectral transmittance of the color filter, which is theoretically $\tau_b(X, Y, \Theta, \Phi; \lambda)$, is independent of geometry and can be written as $\tau_b(\lambda)$. But, if the filter is of uniform thickness (in the Z direction), then a ray at angle of inclination Φ from the Z direction will travel through the filter for a distance proportional to $1/cos\Phi$. Transmittance of light along a path of length p is proportional to τ^p where τ is the transmittance along a path of length 1, and varies with wavelength in general. Thus, $\tau_b(X,Y,\Theta,\Phi;\lambda)$ should properly be rewritten as $\tau_b(\lambda)^{1/cos\Phi}$, which would result in a very complex model when integrated with the geometric reduction of $\boldsymbol{R}$.

Since some of the assumptions are likely true, and some clearly false, we need a criterion for deciding whether to accept a particular assumption for a given system. The appropriate criterion is clearly the degree of conformance of the real system with that assumption. Stated operationally, the criterion should be the magnitude of measurable error introduced by making that assumption for the given system.

We cannot afford to exhaustively calculate or model $\boldsymbol{R}$ and $\boldsymbol{M}$; in fact, if we begin with the most general forms and try to reduce them one assumption at a time, just taking that first step is probably impossible – it would involve measuring very high-dimensional functions with fine precision. Instead, we propose the opposite strategy: Begin with the simplest assumption, and measure how well it predicts the behavior of the real system. If it predicts the real behavior well, then accept the model. If (more likely) it does not, then analyze the observed behavior to see what assumption is most probably the major contributor to the error; re-formulate $\boldsymbol{R}$ and $\boldsymbol{M}$ without relying on that assumption; and test the new formulation of $\boldsymbol{R}$ and $\boldsymbol{M}$. Repeat until done, i.e. until predicted behavior matches observed behavior within the limits mandated by the application system. We call this "top-down" modeling, since we begin with the simplest "high-level" model, and gradually work down through the details to a more realistic model for the system in question.

Two aspects of this process are especially noteworthy:

- How to test a model: When a form is proposed for $\boldsymbol{R}$ and $\boldsymbol{M}$, it will contain a number of constants. These must be estimated by a calibration process in which data is obtained experimentally to determine the optimal values of the constants. Then, based on the residual error of the parameter estimation, or using additional experimental data, the quality of the resulting instantiated model can be determined. Thus, the cycle is: model – measure data – estimate constants – determine error – analyze error – reformulate model – ...
  This cycle shows the integration of calibration, error determination, and model customization, in a single iterative process that is repeated until a satisfactory model and calibration are obtained. It stands in sharp contrast to the traditional process in which equation forms are postulated from first principles, then constants are measured (calibration), and then the process is declared to be finished. In our proposed method, all systems would start with the same simple model, but each system may end up with

its own idiosyncratic form of the model equations, based on observations made with the system as the model is being incrementally refined.

The traditional approach may be adequate for most fixed-parameter camera/lens systems, but for MDOF systems, which are highly individual in behavior, it fails completely as shown in this chapter. The model must be customized for an MDOF system based on its inherent characteristics.

- How to reformulate the model: Traditional optics gives no basis for generalizing the model, since it proposes exact forms for each equation. As we have shown in this chapter, it is frequently necessary to use forms that are more generic, allowing more parameters to be integrated into each function. Of course, such forms involve more constants than simple physical models, and thus require more data measurement. In the limit, a function can be represented as a set of measured values with interpolation, or even as a densely measured table of values. Thus, the process of model generalization proceeds in an orderly succession of forms as follows:
  1. Simple idealized model (0–2 constants),
  2. Parameterized function (few constants),
  3. Generic function, e.g. multivariate polynomial (several constants),
  4. Piecewise generic function, i.e. table + interpolation (many constants) and
  5. Explicit dense table (huge number of constants).

In summary, we see the simple, commonly used equations of imaging as a starting point for MDOF calibration, but assuredly not the final form that will be used. We propose a systematic process of incrementally adding more complexity, modeling power, and experimental data to the process, gradually reducing the discrepancy between the instantiated model and the behavior of the real system, until the model fits closely enough to be accepted for use. The increases in complexity of the models will usually take the form of using a multivariate function to replace two or more functions (i.e. revoking an assumption of separability); to represent the larger function, a more and more generic form, with more calibration constants, may be needed.

Thus, the forms of the equations, as well as the values of the constants in the equations, should result from the calibration process; as opposed to the traditional approach, in which the forms of the equations are assumed and the constants are simply estimated by standard means without further refinement of the equations.

Acknowledgments

This research was sponsored by the Avionics Lab, Wright Research and Development Center, Aeronautical Systems Division (AFSC), US Air Force, Wright-Patterson AFB, OH 45433-6543 under Contract F33615-90-C-1465, ARPA Order No. 7597.

The views and conclusions contained in this document are those of the authors and should not be interpreted as representing the official policies, either expressed or implied, of the US Government.

Reference

1. R. G. Willson. *Modeling and Calibration of Automated Zoom Lenses.* PhD thesis, Carnegie Mellon University, January 1994.

7 System Calibration Through Self-Calibration

Armin Gruen and Horst A. Beyer

Summary

The method of has proved to be one of the most powerful calibration techniques. If used in the context of a general bundle solution it provides for object space coordinates or object features, camera exterior and interior orientation parameters, and models other systematic errors as well. Therefore, because of its flexibility, it may be used in stereo, multi-frame systems, egomotion computations, etc. This chapter gives a brief introduction to the principle of self-calibration, emphasizes some of the problems which are associated with it, and demonstrates with practical data to what extent geometry and network design will influence the determinability of the self-calibration parameters. Finally, a system test will show the high accuracy performance of self-calibrating CCD camera systems.

7.1 Introduction

Since photogrammetry has always been concerned with precise measurements using images, the accurate calibration of the sensors used has been, and still is, of major concern. In the case of a single frame camera sensor (photographic or opto-electronic) the underlying geometrical model for processing is that of perspective projection and the associated procedure for the adjustment of the image coordinate measurements and the estimation of the derived parameters is the "bundle method". The fundamental projection parameters are the image coordinates of the principal point and the camera constant. They define the interior orientation of a CCD frame. According to a widespread definition some authors also include the lens distortion (very often only the radial part) into the set of parameters for interior orientation. These parameters may be systematically distorted. However, there are a great number of additional error sources which may lead to a deformation of the imaging bundle of rays and thus contribute to the overall systematic error budget. Among the most prominent in photographic systems are film and emulsion distortion, unflatness of the imaging plane, false fiducial mark coordinates, effects of image motion, and atmospheric refraction. In CCD camera based systems not all of

Springer Series in Information Sciences, Vol. 34
Calibration and Orientation of Cameras in Computer Vision
Eds.: Gruen, Huang © Springer-Verlag Berlin Heidelberg 2001

those are relevant, but one has to consider a number of additional distortion sources as identified by Gruen [1] and investigated by Beyer [2] and [3].

We define now "system calibration" as a technique which reduces the original image data such that it is most compatible with the chosen parameters of perspective projection. This may include an estimation and correction of these parameters themselves.

In Sect. 7.2 the method of self-calibration, a technique for system calibration which was introduced into photogrammetry thirty years ago, is briefly described.

Self-calibration is a standard procedure in aerial photogrammetry. Under operational conditions it leads to accuracy improvements of up to a factor 3. It is of particular value in CCD camera based close-range systems since the cameras and framegrabbers used are commonly poorly calibrated or the calibration parameters might significantly change over time (camera constant and lens distortion through focussing, principal point through warm-up effects, etc.). Accuracy improvement factors of up to 10 were observed in controlled test projects by Gruen et al. [4] and by Beyer [2,3]. Since self-calibration adds new parameters to the linearized system of bundle adjustment there is a certain danger of overparameterization, which could lead to ill-conditioning or even singularity of the normal equations of least squares adjustment.

Section 7.3 demonstrates how geometry and network design influence the determinability of the self-calibration parameters.

Section 7.4 shows the potential of this technique in a controlled test and proves the high accuracy performance of CCD camera based systems.

7.2   The Concept of Self-Calibration

If single-frame camera data is processed, for instance in CCD camera applications, the geometric sensor model is that of perspective projection, leading to the so-called "bundle method". This bundle method is considered the most flexible, general and accurate sensor model. Long before it became a standard procedure in aerial photogrammetry it was used in a variety of close-range applications. Since, for the method of self-calibration presented here, the bundle method is the underlying estimation model the latter will be very briefly addressed in the following (there exists an abundance of publications on this subject in the photogrammetric literature, e.g. [5–9]).

7.2.1   The Bundle Method

The basis of the bundle method is the collinearity condition (compare also Fig. 7.1)

$$\begin{bmatrix} X \\ Y \\ Z \end{bmatrix}_i = \lambda_{ij} \boldsymbol{R}_j \begin{bmatrix} x_{ij} - x_{0j} \\ y_{ij} - y_{0j} \\ 0 - c_j \end{bmatrix} + \begin{bmatrix} X_0 \\ Y_0 \\ Z_0 \end{bmatrix}_j \qquad (7.1)$$

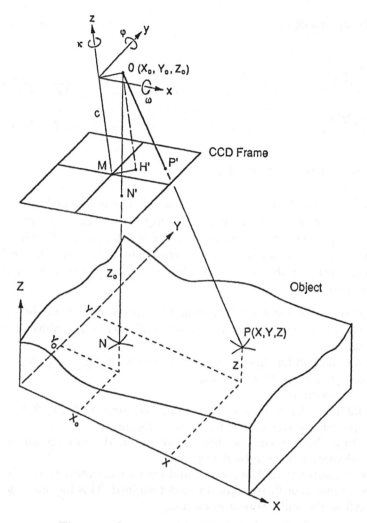

Fig. 7.1. Sensor model of the bundle method.

where X_i, Y_i, Z_i are the object space coordinates of object point (P_i); $X_{0j}, Y_{0j},$ Z_{0j} are the object space coordinates of perspective center (O_j); $x_{ij}, y_{ij}, 0$ are the measured image coordinates of point (P'_{ij}); $x_{0j}, y_{0j},$ are the image space coordinates of principal point (H'_j); c_j is the camera constant of CCD frame j; R_j is the rotation matrix (orthogonal) between image and object space coordinate systems; λ_{ij} is the scale factor for imaging ray ij; and $i = 1, \ldots,$ nop (nop = number of object points), $j = 1, \ldots,$ nof (nof = number of CCD frames).

In (7.1) the interior orientation of a CCD frame j is defined by the parameters x_{0j}, y_{0j}, c_j while the parameters $X_{0j}, Y_{0j}, Z_{0j}, R_j (\varphi_j, \omega_j, \kappa_j)$ define the exterior orientation. Here $\varphi_j, \omega_j, \kappa_j$ are the three rotation angles which build up the rotation matrix R_j. The three components in (7.1) are reduced to two by cancelling out the scale factor λ_{ij}, and then rearranged according to

$$x_{ij} = -c_j f_{ij}^x + x_{0j} = -c_j \frac{r_{11j}(X_i - X_{0j}) + r_{21j}(Y_i - Y_{0j}) + r_{31j}(Z_i - Z_{0j})}{r_{13j}(X_i - X_{0j}) + r_{23j}(Y_i - Y_{0j}) + r_{33j}(Z_i - Z_{0j})} + x_{0j}$$

$$y_{ij} = -c_j f_{ij}^y + y_{0j} = -c_j \frac{r_{12j}(X_i - X_{0j}) + r_{22j}(Y_i - Y_{0j}) + r_{32j}(Z_i - Z_{0j})}{r_{13j}(X_i - X_{0j}) + r_{23j}(Y_i - Y_{0j}) + r_{33j}(Z_i - Z_{0j})} + y_{0j}$$

$$(7.2)$$

$r_{11j} \ldots r_{33j}$ are the elements of $\boldsymbol{R}_j$.

Since (7.2) provide for a general sensor model they accommodate easily the specific cases which are often treated in computer vision, e.g. stereo, egomotion, etc. Depending on the parameters which are considered either known a priori or treated as unknowns these equations may result in the following cases, which are also listed in Table 7.1 (x_{ij}, y_{ij} are always regarded as observed quantities):

(a) General bundle method: All parameters on the right-hand side of (7.2) are unknown (interior orientation, exterior orientation, object point coordinates).
(b) Bundle method for "metric camera" systems: x_{0j}, y_{0j}, c_j (interior orientation) are given, all others unknown.
(c) Spatial resection:
  (i) Interior orientation and object point coordinates (X_i, Y_i, Z_i) are given, the exterior orientation has to be determined.
  (ii) Only object point coordinates are given, the interior and exterior orientation have to be determined.
(d) Spatial intersection: The interior and exterior orientation are given, the object point coordinates have to be determined. This includes the stereo as well as the multiframe approaches.

Any combination of procedures is possible within the general bundle concept. Also incomplete parameter sets of exterior/interior orientation and object points can be treated.

7.2.2 Least Squares Estimation

Equations (7.2) are considered observation equations functionally relating the observations x_{ij}, y_{ij} to the parameters of the right-hand side according to

$$l = f(\boldsymbol{x}). \tag{7.3}$$

For the estimation of $\boldsymbol{x}$ the Gauss-Markov model of least squares is used. After linearization of (7.3) and the introduction of a true error vector $\boldsymbol{e}$ we obtain

$$l - \boldsymbol{e} = \boldsymbol{A}\boldsymbol{x}. \tag{7.4}$$

Table 7.1. Photogrammetric orientation and point positioning procedures as special cases of the general bundle method.

Procedure	Given parameters	Unknown parameters
General bundle		$(X,Y,Z)_i; IO_j; EO_j$
Metric camera bundle	IO_j	$(X,Y,Z)_i; EO_j$
Spatial resection (a)	$IO_j; (X,Y,Z)_i$	EO_j
(b)	$(X,Y,Z)_i$	$IO_j; EO_j$
Spatial intersection (stereo or multiframe)	$IO_j; EO_j$	$(X,Y,Z)_i$

$IO\ldots$ Interior orientation, $EO\ldots$ Exterior orientation

The design matrix $\boldsymbol{A}$ is a $n \times u$ matrix (n is the number of observations, u is the number of unknown parameters), with $n \geq u$ and, in general, rank $(\boldsymbol{A}) = u$.

With the assumed expectation $E(e) = \mathbf{0}$ and the dispersion operator D we get

$$E(\boldsymbol{l}) = \boldsymbol{Ax} ,$$
$$D(\boldsymbol{l}) = \boldsymbol{C}_{ll} = \sigma_0^2 \boldsymbol{P}^{-1},$$
$$D(\boldsymbol{e}) = \boldsymbol{C}_{ee} = \boldsymbol{C}_{ll} .$$
(7.5)

$\boldsymbol{P}$ is the "weight coefficient" matrix of $\boldsymbol{l}$ and $\boldsymbol{C}$ stands for covariance operator; σ_0^2 is the variance factor.

The estimation of $\boldsymbol{x}$ and σ_0^2 is usually (not exclusively) attempted as unbiased, minimum variance estimation, performed by means of least squares, and results in:

$$\hat{\boldsymbol{x}} = \left(\boldsymbol{A}^{\mathrm{T}} \boldsymbol{PA}\right)^{-1} \boldsymbol{A}^{\mathrm{T}} \boldsymbol{Pl} ,$$
$$\boldsymbol{v} = \boldsymbol{A}\hat{\boldsymbol{x}} - \boldsymbol{l} ,$$
$$\hat{\sigma}_0^2 = \frac{\boldsymbol{v}^{\mathrm{T}} \boldsymbol{Pv}}{r}, \quad r = n - u .$$
(7.6)

where $\boldsymbol{v}$ are the residuals of least squares.

The structure of $\boldsymbol{A}$ is determined by the procedure applied (compare Table 7.1), it reflects the overall network design and thus also the geometrical and numerical stability of the arrangement.

For $\boldsymbol{A}^{\mathrm{T}} \boldsymbol{PA}$ to be uniquely invertible, as required in (7.6), the network needs an external "datum", that is, the seven parameters of a spatial similarity transformation of the network need to be fixed. This is usually achieved either by introducing control points with seven fixed coordinate values, or by fixing seven appropriate elements of the exterior orientations of two frames. The precision of the parameter vector $\boldsymbol{x}$ is controlled by its covariance matrix $\boldsymbol{C}_{xx} = \hat{\sigma}_0^2 \left(\boldsymbol{A}^{\mathrm{T}} \boldsymbol{PA}\right)^{-1}.$

7.2.3   Systematic Error Compensation by Self-Calibration

For systematic error compensation a number of methods have evolved over the years, among which the following are the most prominent:

A priori: • Data reduction using physical models
                     • Laboratory calibration
                     • Réseau corrections
                     • Testfield calibration under project conditions

A posteriori: • Post-treatment of adjustment results (analysis of residuals)

Simultaneous: • Compensation by network arrangement
                     • Self-calibration by additional parameters

In the following only the method of self-calibration will be addressed because it has proved to be the most flexible and powerful technique, while requiring at the same time no additional measurement effort.

Self-calibration by additional parameters essentially consists in expanding the right-hand side of (7.2) by additional functions which are supposed to model the systematic image errors according to

$$x_{ij} = -c_j f_{ij}^x + x_{0j} + \Delta x_{ij} \,,$$
$$y_{ij} = -c_j f_{ij}^y + y_{0j} + \Delta y_{ij} \,. \tag{7.7}$$

The terms $\Delta x_{ij}, \Delta y_{ij}$ can be understood as corrections to the image coordinates x_{ij}, y_{ij} in order to reduce the physical reality of the sensor geometry to the perspective model. Obviously the principal point coordinates x_{0j}, y_{0j} can be considered part of this correction term. If a parameter for the camera constant is added, the full interior orientation is included in the additional parameter set.

The formulation of the additional parameter function is very crucial for a successful self-calibration. The actual physically caused deformations, which are a priori unknown in structure and magnitude, should be modeled as closely and as completely as possible. On the other hand, the chosen parameter set must be safely determinable in a given network arrangement. These are essentially two conflicting aspects which will be addressed in Sect. 7.2.4.

In aerial photogrammetry the sources for systematic errors have been studied in great detail and are generally quite well understood and modeled [9]. An international test has shown that different modeling concepts lead to practically the same results as long as the systematic errors are fully covered in the respective models [10].

In photographic close-range systems the following functions have proved to be effective [11]:

$$\Delta x = -\Delta x_0 + \frac{\bar{x}}{c}\Delta c + \bar{x}s_x + \bar{y}a + \bar{x}r^2 k_1 + \bar{x}r^4 k_2 + \bar{x}r^6 k_3$$
$$+ \left(r^2 + 2\bar{x}^2\right)p_1 + 2\bar{x}\bar{y}p_2 ,$$

$$\Delta y = -\Delta y_0 + \frac{\bar{y}}{c}\Delta c \;\; + \;\; 0 \;\; + \;\; \bar{x}a + \bar{y}r^2 k_1 + \bar{y}r^4 k_2 + \bar{y}r^6 k_3$$
$$+ 2\bar{x}\bar{y}p_1 + \left(r^2 + 2\bar{y}^2\right)p_2 , \qquad (7.8)$$

with $\bar{x} = x - x_0, \qquad \bar{y} = y - y_0, \qquad r^2 = \bar{x}^2 + \bar{y}^2$

(the indices $_{ij}$ are left out here for the sake of simplicity).

This is called a "physical model", because all its components can directly be attributed to physical error sources. The individual parameters represent:

$\Delta x_0, \Delta y_0, \Delta c$ ... change in interior orientation elements

s_x .... scale factor in x ("affinity")

a .... shear factor (jointly in x, y)

$k_1, k_2, k_3,$ ...... first three parameters of radial symmetric lens distortion (k_3 is a priori disregarded if normal- and wide-angle lenses are used)

p_1, p_2 .... first two parameters of decentering distortion

   Equations (7.8) have also proved to be successful in CCD camera systems [2,3] and will be used in the investigations of this chapter.

   The location of the principal point is not specified for most CCD cameras, varies from camera to camera and depends on the configuration of the frame grabber. The scale factor in x is required to model the imprecise specification of the sensor element spacing and additonal imprecisions introduced with PLL line-synchronization. In the latter case the pixel spacing in x must be computed from the sensor element spacing, the sensor clock frequency and the sampling frequency with:

$$psx = ssx \frac{f_{\text{sensor}}}{f_{\text{sampling}}} \qquad (7.9)$$

where psx is the pixel spacing in x, ssx is the sensor element spacing in x, f_{sensor} is the sensor clock frequency, f_{sampling} is the sampling frequency of the frame grabber.

   The shear factor a must be included to compensate for the geometric deformation which can be induced by PLL line-synchronization [3]. The use of additional parameters leads to an extended bundle model

$$l - e = Ax + A_3 z. \qquad (7.10)$$

where z, A_3 are the vector of additional parameters and the associated design matrix.

In a general bundle concept all unknown parameters are treated as stochastic variables. This permits us to consider a priori information about these parameters to be included and includes both extreme cases where parameters are either excluded from the model or may be treated as free unknowns. If we split the vector x into its components x_p (for object point coordinates) and t (for exterior orientation parameters) we obtain the following estimation model:

$$-e_B = A_1 x_p + A_2 t + A_3 z - l_B \; ; \; P_B$$

$$-e_p = \quad I\, x_p \qquad\qquad - l_p \; ; \; P_p$$

$$-e_t = \qquad\qquad I\, t \qquad - l_t \; ; \; P_t \qquad\qquad (7.11)$$

$$-e_z = \qquad\qquad\qquad I\, z - l_z \; ; \; P_z$$

where e_B, e_p, e_t, e_z are vectors of true errors of image coordinates, object point coordinates, exterior orientation elements, additional parameters, $l_B, l_p,$ l_t, l_z are vectors of observations of image coordinates (minus constant term from Taylor expansion), object point coordinates, exterior orientation elements, additional parameters, P_B, P_p, P_t, P_z are associated weight coefficient matrices, $x_p, t, z,$ are parameter vectors of object point coordinates, exterior orientation elements, additional parameters, A_1, A_2, A_3 are associated design matrices, and I is the Identity matrix.

In the investigations of Sect. 7.3 we will treat all control point coordinates with infinite weight (diag $(P_p) \to \infty$), that is error-free and with $l_p = 0$ (which leads to $x_p = 0$, as x_p does not represent the full coordinate values but because of previous linearization only the incremental corrections). All other object points will be considered as "new points" with $P_p = 0$ and $l_p = 0$. The same will apply to all exterior orientatiom elements: $P_t = 0$, $l_t = 0$. In some cases the additional parameters will also be assumed as free unknowns, with $P_z = 0$, $l_z = 0$. The stabilizing effect of finite (nonzero) weights for additional parameters in the case of weak determinability was shown by Gruen [7,9] and will also be addressed below.

7.2.4   Treatment of Additional Parameters

In (7.11) a unique set of additional parameters may be introduced for all participating frames or, as necessary if each frame comes from a different CCD camera, an individual set may be assigned to each frame. Especially in the latter case the danger of overparameterization becomes apparent. This is the reason for the need to have a powerful parameter control procedure available, for the automatic detection of nondeterminable parameters and their exclusion from the system.

In the past such procedures have been developed by different researchers. They all rely on statistical concepts. The procedures have never been com-

pared to each other in an independent test. The first author of this chapter
has published his approach in a series of papers [7,9,12,13]. In the following
a summary of this procedure will be given.

Conditions for the algebraic determinability of additional parameters were
previously formulated by Gruen [7] and are reiterated in Appendix A. Since it
is necessary in practice to deal with erroneous observations, the purely algebraic approach must be given up in favor of a statistical approach. Therefore
any decisions will be correct only with a certain probability. The major problem consists in finding criteria for the rejection of individual parameters. The
checking of parameters must be related to the purpose of the triangulation.
Here three different objectives must be distinguished:

(a) point positioning
(b) optimum estimation of the elements of exterior orientation, e.g. for ego-
    motion purposes
(c) analysis of systematic image errors aimed at the optimum estimation of
    these errors.

Objective (a) involves the optimum estimation of object space coordinates.
Additional parameters are used as supporting variables for the improvement
of the estimation model; they do not have a separate, independent meaning. Hence their statistical significance is not a matter of concern, unless
insignificant parameters cause a substantial decrease of the redundancy in
small systems. The precision measures of the object space coordinates are the
system's only available meaningful parameters for quality control. In other
words, those additional parameters that would cause an inadmissibly large
deterioration of the network's precision measures have to be excluded. The
most popular precision measures are the mean and maximal variance of the
object point coordinates. It is necessary to check to what extent those measures are deteriorated by certain additional parameters.

Since the determinability of additional parameters may vary largely, their
checking should be done at different stages of the least squares adjustment
process, using rejection criteria at varying sensitivity levels. The following
stepwise procedure can be operated in a reasonably fast mode.

(1) In order to avoid a quasi-column rank deficiency in the design matrix
    of the estimation model, the additional parameters are introduced as
    observed variables. The assignment of small weights assures that only a
    very small constraint is applied.
(2) Very poorly determined additional parameters can be deleted in the
    course of the factorization procedure of the normal equations. Weakness
    of a particular additional parameter is indicated by a comparable small
    pivot element. If such a pivot falls short of a certain limit, the related
    additional parameter needs to be deleted. This can be achieved by adding
    a large number to this pivot element. The reduction process is continued.

(3) In a correlation check, high correlations indicate an inherent weakness of the system. They are particularly damaging if they occur between additional parameters and object point coordinates. Any additional parameter which leads to such correlations larger than 0.9 should be rejected. For computational reasons, only a few object points in characteristic locations may be included here.

(4) Detection and deletion are effected by those additional parameters that belong to the critical range between poorly determined and sufficiently well determined parameters. The trace check of the covariance matrix (outlined in Appendix B) may be used here.

(5) In a final step, the remaining additional parameters can be tested for significance, if this is considered necessary. It is suggested that the a posteriori orthogonalized additional parameter vector should be tested following the method outlined by Gruen [13]. The nonsignificant orthogonal components are set to zero. The back-transformation of the thus modified additional parameter vector from the orthogonal space to the "regular" space provides for a cleaned additional parameter set, involving the significant part only.

7.3 Determinability of Self-Calibration Parameters Under Various Network Conditions

In the following we will show with the help of practical data to what extent individual or sets of additional parameters (APs) can be determined under varying network conditions. Our 3-D testfield is used as the object to be measured.

Figure 7.2 shows schematically the object and the arrangement of CCD camera stations. For image acquisition a SONY-XC77CE camera with a 9 mm lens was used. The imagery was acquired with a VideoPix frame grabber using PLL line-synchronization. The resultant imagery has a size of 768×575 pixels. The sampling rate of the frame grabber was approximately 14 MHz, resulting in a pixel spacing of about 11µm. A total of 36 object points were measured, 24 on the wall and 12 on the structure. Figure 7.3 shows a typical CCD image. The average number of measured image points per frame is 31.

The following parameters will be varied in our computational versions:

- Number of frames used in bundle adjustment (1, 2, 3, 4, 8).
- Configuration of frames (small base – parallel optical axes; large base – convergent optical axes; additional rotation of frames by 90° at the approximate position of the nonrotated frames)
- Object depth (plane object – rear wall, two depth planes – rear wall and three rods in the foreground)

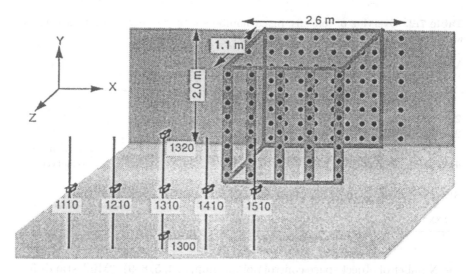

Fig. 7.2. View of the test arrangement.

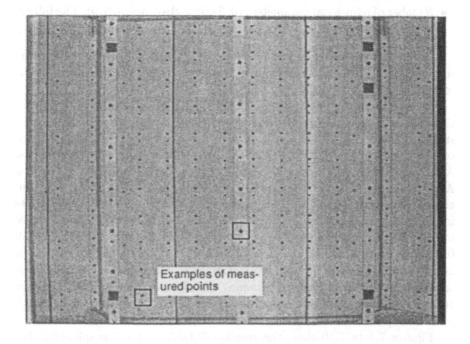

Fig. 7.3. Image grabbed at station 1310. The average number of measured points is 31 per image from a total of 36 object points.

Table 7.2. Overview of versions. Image numbers with xxx1 denote that the camera was rotated by 90° around its optical axis.

Versions	Frames	Config- uration	Images	APs
1xxxx	1		1310	
21xxxx	2	1	1210, 1410	
22xxxx	2	2	1110 1510	
3xxxx	3		1110 1310,1510	$\Delta x_H, \Delta y_H, \Delta c$
41xxxx	4	1	1110,1300,1320,1510	observed
42xxxx	4	2	1110,1111, 1510, 1511	
81xxxx	8	1	1110,1111,1300,1301,1320,1321, 1510,1511	
82xxxx	8	2	1110,1111,1300,1301,1320,1321,1510,1511	all free

- Number of object space control points (min, 3, 4, 5, 8, 9). "Min" stands for seven control coordinates, the minimum number for all versions including more than one frame.
- Number of additional parameters $(0, 3\,[\Delta x_0, \Delta y_0, \Delta c]\,, 4, 5, 9)$. The full set of nine APs is given by (7.8) leaving out k_3.

Table 7.2 together with Fig. 7.2 shows the frame and configuration versions together with an indication as to how the APs are treated a priori (observed or free).

The results of computations are listed in Appendix C, Tables C1–C7. Besides the estimated accuracy of the image coordinates $(\hat{\sigma}_0)$ and the average standard deviations of the object points $(\bar{\sigma}_X, \bar{\sigma}_Y, \bar{\sigma}_Z, \text{or } \bar{\sigma}_{XYZ})$ the maximum correlations are listed (AP-OP: between APs and object point coordinates, AP-EO: between APs and exterior orientation elements, AP-AP: between APs only). The inherent correlation between k_1 and k_2 ($> 90\%$) is not included. The control point versions are selected according to the use of targets in one and two planes, i.e. for one plane: min, 3, 4, 5, 9 and for two planes: min, 4, 5, 8 control points. The control points have standard deviations of 0.0 (1.e-31) mm in X, Y, and Z. Standard deviation of a priori unit weight is 1.1 micrometer or 1/10th of the pixel spacing. The image coordinates have weight 1. $\Delta x_0, \Delta y_0, \Delta c$ have a standard deviation of 0.1 mm when they are treated as observed (in some cases, which are marked, one of 1.0 or 0.01). All additional parameters are treated as free unknowns for versions 82xxxx.

Figures 7.4 to 7.10 show the results of Tables C1–C7 graphically. The average standard deviation of all object coordinates $(\bar{\sigma}_{XYZ})$ is plotted against the number of additional parameters (APs) in order to show the influence of the latter on the former in a particular network version. The following symbols are used throughout these figures:

———————   Two object planes

- - - - - - -   One object plane

$\boxed{\text{M}}$ Number of control points; M stands for "minimum"

As a general result, with increasing number of frames, larger bases and convergent optical axes, more explicit 3-D object extension and increasing number of control points, we get an improvement in the determinability of APs. Since a detailed analysis would require much space, we will restrict our comments in the following to the most prominent points. The analysis will be based solely on the evaluation of $\bar{\sigma}_X, \bar{\sigma}_Y, \bar{\sigma}_Z,$ or $\bar{\sigma}_{XYZ}$ respectively. Any other aspects, like the amount of correlation which a certain AP generates, are not considered.

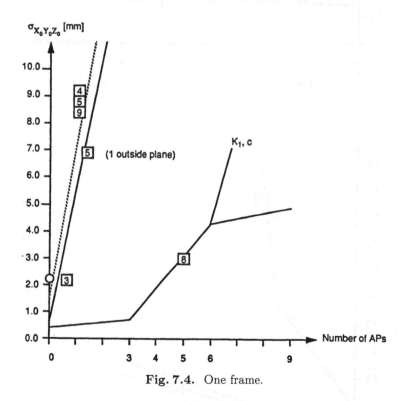

Fig. 7.4. One frame.

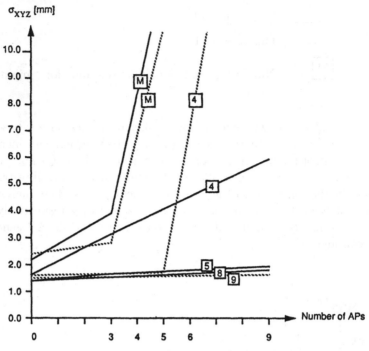

Fig. 8.17. Two frames, configuration 1.

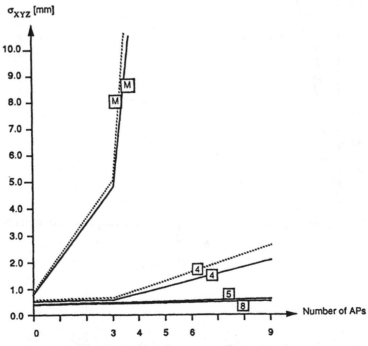

Fig. 8.18. Two frames, configuration 2.

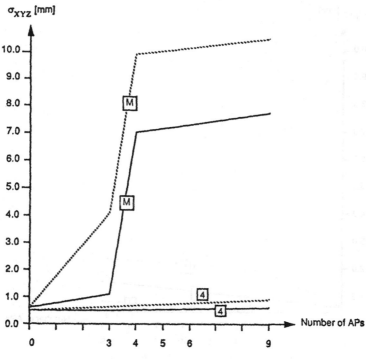

Fig. 7.7. Three frames.

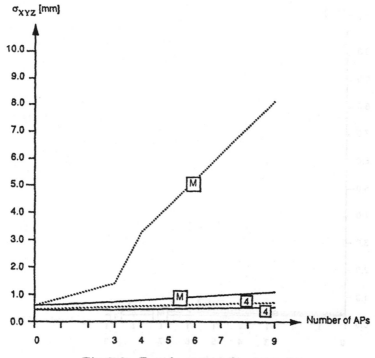

Fig. 7.8. Four frames, configuration 1.

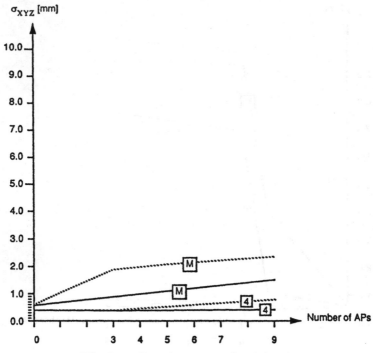

Fig. 7.9. Four frames, configuration 2.

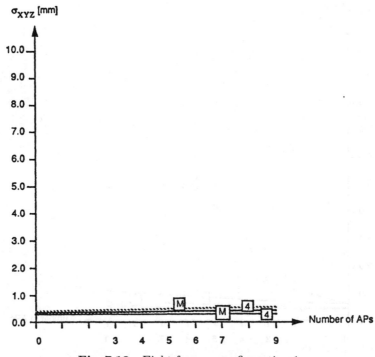

Fig. 7.10. Eight frames, configuration 1.

One frame (spatial resection): The elements of interior orientation cannot be determined as long as a plane object is used, independent of the number of control points. Other APs, however, might be determinable, given a sufficient number and distribution of control points. In the event of a substantial 3-D object extension the interior orientation is determinable, assuming at least five control points in appropriate 3-D distribution (here: three in one depth plane, two in the other). Attention: In version 12050, $P = 2$, $Co = 5$ only one control point has been used outside the rear plane, therefore all three interior orientation elements cannot be determined simultaneously (this one control point provides for only two image coordinate observations, whereas three parameters are to be determined).

Two frames (stereo, small base): Here some of the results look erratic. The versions with one object plane ($P = 1$) produce partly better object point coordinate standard deviations if a larger a priori standard deviation for the interior orientation parameters is used (1 mm versus 0.1 mm). Also the values are partly too good to be possibly correct. This is due to the fact that all versions P = 1, AP = 3 are highly unstable (correlations AP-EO around 100%) so that any kind of numerical distortion may show up. It must be emphasized that in these computations the APs were not tested according to the procedure suggested in Sect. 7.2.4 and Appendix A, therefore highly instable APs were not automatically deleted prior to further analysis. Only with two object planes ($P = 2$) and five control points ($Co = 5$) or more can we get the interior orientation parameters (and other APs) accurately enough.

Two frames (convergent, large base): The previous situation changes in parts drastically if a larger base and convergent frames are used. Still in the case of minimum control the principal point coordinates $\Delta x_0, \Delta y_0$ are not determinable, but this is corrected by an additional control point. For $P = 2$ and five control points all nine APs are determined very well.

Three frames (large base plus center frame): As can be expected, this version gives better results than the previous version. While we still have problems with the minimum control versions, four control points give quite good results, especially in the two object planes ($P = 2$) case.

Four frames (large horizontal base plus vertical base): Compared to the previous version we now obtain even in the case of minimum control/two planes good results for the interior orientation parameters and decent results for all nine APs. The corresponding one plane version is still weak in the interior orientation parameters and unacceptable for the other APs.

Four frames (large horizontal base plus additional 90% rotation of frames): As in the previous version, the four control point cases give good results up to all nine APs. However, the two plane/minimum control version is slightly worse, whereas the one plane/minimum control version gives substantially better results for nine APs than before, although it cannot be considered good enough yet.

Eight frames (large horizontal base plus vertical base plus additional 90% rotation of frames): Here we find excellent results in all versions, even for minimum control and only one object plane. Also, the use of a priori unconstrained APs does not worsen the results, as evidenced by the results of Table 7.3.

Summarizing the results, only the eight frames version was capable of producing very good results under all conditions assumed. Releasing either the nonredundant datum (minimum control) condition or the 3-D object condition even the four frames (two bases) arrangement may produce acceptable results. If only three frames are used, aligned along a common base, decent results can only be achieved with redundant control information ($\geq$ four control points). On the other hand, the results also show the inherent weakness of a two frames (stereo) system for self-calibration. It must be emphasized that these results have been obtained by using block-invariant additional parameters, that is one set of coefficients for all frames. If in robotics for example several CCD cameras are utilized simultaneously and if problems, such as on-line focussing, zooming and compensation of damage caused by physical disturbances (vibrations, shocks, electrical distortions, etc.) have to be handled, a frame-invariant approach might be necessary and lead possibly to totally different conclusions than those presented here. In any case it would require a much more constrained environment. No matter what the actual approach is, a "blind" use of additional parameters is never recommended. In a concrete environment the geometrical conditions might be so complex that it is very difficult to predict the results of self-calibration. Therefore the use of additional parameters should always be accompanied by a sophisticated checking and statistical testing procedure.

7.4   A Final System Test

This is to demonstrate the accuracy potential of CCD camera based systems if the calibration is done properly. The version of eight frames (large bases, convergent optical axes, 90° rotated frames, a priori unconstrained APs) was used with two control point versions (82200X: Minimum – seven coordinates, 82205X: Five full control points). The average standard deviation of image coordinates from least squares matching is 1.64 and 0.74 μm in x and y respectively. The targets are imaged on the average in 6.9 frames of the 8 frames used in the accuracy test. The image scale is 1 : 352. The targets are imaged on average with a diameter of 5.2 pixels, with a minimum of 3.5 and a maximum of 7.4. The average object distance is 3233 mm. The depth of the object is thus approximately 1/3 of the average object distance. Table 7.3 shows the results of computations.

Table 7.3. Results of bundle adjustment. A priori standard deviations for additional parameters are infinity ($\hat\sigma_X, \hat\sigma_Y, \hat\sigma_Z$ were computed with $\sigma_0 = 1.1$ μm).

Version	AP	Co	Ch	r	$\hat\sigma_0$ [μm]	$\bar\sigma_X$ [mm]	$\bar\sigma_Y$ [mm]	$\bar\sigma_Z$ [mm]	μ_X [mm]	μ_Y [mm]	μ_Z [mm]	μ_x [μm]	μ_y [μm]
822000	0	min	33	351	7.94	0.359	0.443	0.574	4.404	4.161	3.849	12.43	12.08
822009	9	min	33	342	1.53	0.382	0.462	0.574	0.338	0.325	0.529	1.09	1.05
822050	0	5	31	359	8.44	0.200	0.199	0.425	4.147	3.906	5.501	12.57	12.20
822059	9	5	31	350	1.52	0.221	0.225	0.427	0.306	0.314	0.503	1.05	1.02
Improvement 822000 / 822009					5.2	0.94	0.96	1.0	13.0	12.8	11.4	11.4	11.5
Improvement 822005 / 822059					5.6	0.90	0.88	1.0	13.6	12.4	10.9	12.0	12.0

AP .... Number of additional parameters

Co .... Number of control points

Ch .... Number of check points

r .... Redundancy

$\hat\sigma_0$ .... Standard deviation of unit weight a posteriori; this corresponds to the estimated standard deviation of image coordinates

$\bar\sigma_X, \bar\sigma_Y, \bar\sigma_Z$ ...... Theoretical precision values of check point coordinates

μ_X, μ_Y, μ_Z ...... root mean square errors from comparison of estimated coordinates to check point coordinates in object space

μ_x, μ_y .... root mean square errors from comparison of estimated coordinates to check point coordinates in image space

Remark: The a priori weights of additional parameters are 0. For the computations of $\mu_X, \mu_Y, \mu_Z, \mu_x, \mu_y$ a 3-D similarity transformation onto all check points was performed before the comparison

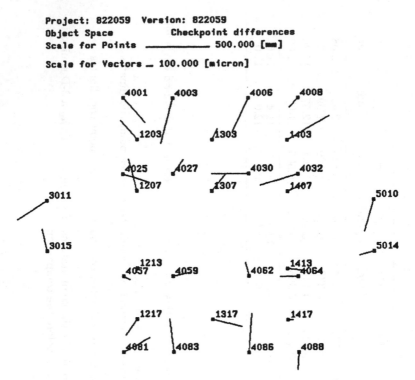

Fig. 7.11. Checkpoint residuals in XY-plane of object space (version 822059).

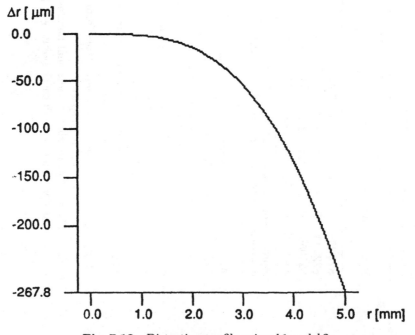

Fig. 7.12. Distortion profile using $k1$ and $k2$.

Raster spacing: 1.0 mm

Vectors ——— 100 µm

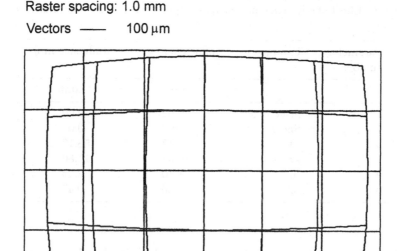

Fig. 7.13. Influence of additional parameters on grid.

The empirical RMSE (μ_X, μ_Y, μ_Z) agrees well with the average standard deviations of the check point coordinates in the self-calibrating versions. The average accuracy improvement through self-calibration is around factor 12. Figure 7.11 shows the distribution of the check point residuals in the XY-plane. Globally this distribution looks rather random.

When analysing these results one should clearly keep in mind that the last measurement of the testfield's reference coordinates dates back 12 months and that at the time of comparison we had no indication of the quality of those coordinates. Figures 7.12 and 7.13 indicate the large effect of the radial symmetric distortion and the lesser influence of the other APs. At the corners of the CCD chip the radial symmetric distortion produces a displacement of 7.3 pixels. This large deformation is easily visible in the CCD frames (compare the bent rods in Fig. 7.3). Table 7.4 gives the effect of the APs at a point with image coordinates $x = 3$ mm, $y = 2$ mm. It is evident that the major source of deformation is the radial symmetric distortion.

7.5   Conclusions

In the context of bundle adjustment the method of self-calibration presents a powerful tool for calibration and systematic error compensation in CCD camera based vision systems. Moreover, it provides for accurate orientation and location of the sensor (spatial resection, egomotion) and for accurate reconstruction of the object space. As a prerequisite, however, the proper

Table 7.4. Effect of additional parameters on point with image coordinates $x = 3$, $y = 2$ mm.

Parameter number	Parameter	Influence in x	Influence in y
		[μm]	[μm]
4	Scale in x	0.02	0.00
5	Shear	−0.00	−0.01
6	k1	−81.47	−58.18
7	k2	16.52	11.80
9	p1	1.10	0.45
10	p2	−0.26	0.47

functions for modeling the systematic errors have to be chosen and a sophisticated checking and testing procedure for the additional parameters has to be incorporated. Furthermore, in order to provide for stable additional parameters and good determinability some geometrical conditions should be observed, like the use of more than two CCD frames with fairly large bases and convergent optical axes. Also, a 3-D object is to be preferred over a 2-D distribution of object points. These geometrical conditions may conflict with the requirements for successful image matching and image tracking in sequences (small disparities, "smooth" object, no occlusions, etc.), but a compromise should be aspired for.

As shown in this chapter and in previous publications, even under relatively weak external constraints ($\leq$ five control points, eight CCD frames, standard set of block-invariant additional parameters, PLL line-synchronization) an accuracy of 1/10th of a pixel or better for the X, Y coordinates of well-defined object points and a depth accuracy of 1/10 000 of the average object distance can be achieved.

Appendix A
Algebraic Determinability of Additional Parameters

An observation vector l_{RS} may include random and systematic components, i.e. a true random error e_R and a true systematic error e_S. Then we obtain

$$e_R + e_S = l_{RS} - E(l_R) = l_{RS} - Ax, \tag{A1}$$

and the estimators for x and/or σ_0^2

$$\bar{x} = \left(A^\mathrm{T}PA\right)^{-1} A^\mathrm{T}Pl_{RS},$$

$$\bar{\sigma}_0^2 = \frac{v_{RS}^\mathrm{T} Pv_{RS}}{r} \tag{A2}$$

are no longer unbiased, if the systematic errors are not modeled and determinable by additional parameters.

We get the residuals v_{RS} to

$$v_{RS} = -\left(I - A\left(A^{\mathrm{T}}PA\right)^{-1}A^{\mathrm{T}}P\right)l_{RS},$$

(A3)

$$v_{RS} = -\left(I - A\left(A^{\mathrm{T}}PA\right)^{-1}A^{\mathrm{T}}P\right)(e_R + e_S)$$
$$= -(I - K)(e_R + e_S) = Me_R + Me_S.$$

v_{RS} is but a visible component of the random and the systematic errors. The systematic component

$$v_S = Me_S,$$

(A4)

plays an important role in determination problems of systematic errors. A systematic error is undeterminable if

$$v_S = Me_S = (K - I)e_S = 0.$$

(A5)

In this case even additional parameters cannot provide for a determination.

Proof: Let $A_1 x_1 = l - e_S$; P be the linear system for the estimation of x_1 and e_S the true systematic error vector. For the residual vector v_S we obtain

$$v_S = -\left(I - A_1\left(A_1^{\mathrm{T}}PA_1\right)^{-1}A_1^{\mathrm{T}}P\right)e_S = Me_S.$$

Let $A_1 x_1 + A_2 x_2 = l - e_S$ and P be the same linear system, extended by the additional parameter function $A_2 x_2$ which describes the systematic error e_S. Then the corresponding normal equations result in

$$\begin{pmatrix} A_1^{\mathrm{T}}PA_1 & A_1^{\mathrm{T}}PA_2 \\ A_2^{\mathrm{T}}PA_1 & A_2^{\mathrm{T}}PA_2 \end{pmatrix}\begin{pmatrix} \hat{x}_1 \\ \hat{x}_2 \end{pmatrix} = \begin{pmatrix} A_1^{\mathrm{T}}Pl \\ A_2^{\mathrm{T}}Pl \end{pmatrix},$$

and for the estimator $\hat{x}_2$ we obtain

$$L\hat{x}_2 = \left(A_2^{\mathrm{T}}PA_2 - A_2^{\mathrm{T}}PA_1\left(A_1^{\mathrm{T}}PA_1\right)^{-1}A_1^{\mathrm{T}}PA_2\right)\hat{x}_2$$

$$= A_2^{\mathrm{T}}P\left(I - A_1\left(A_1^{\mathrm{T}}PA_1\right)^{-1}A_1^{\mathrm{T}}P\right)l = -A_2^{\mathrm{T}}Pv_S.$$

If $v_S = 0$ we get the homogeneous linear system $L\hat{x}_2 = 0$. The non-trivial solutions $\hat{x}_2 \neq 0$ are only existent if L is singular, which however implies the impossibility of obtaining a unique solution $\hat{x}_2 \neq 0$ for x_2.

A reformulation of (A5) yields

$$Me_S = (K - \lambda I)\, e_S = 0\,, \text{ with } \lambda = 1\,. \tag{A6}$$

Because M is singular, $\det (M) = 0$ is always valid, so that other solutions than $e_S = 0$ exist. The solution vectors for e_S can be interpreted as the eigenvectors for the eigenvalue $\lambda = 1$ of K. The existence of $\lambda = 1$ follows from the idempotency of K and the rank deficiency $d_K \neq n$.

There may even exist more than one linearly independent eigenvectors; their number is $u\,(u = d_M = \text{rank deficit of } M = \text{rank}(K)$, if P is regular), because u-fold eigenvalues $\lambda = 1$ are possible.

All this means that for every bundle system u independent vectors of systematic errors exist, which are not determinable. It becomes an interesting and essential question whether this mathematical set of errors includes some vectors which are physically possible, and moreover, which do occasionally or even often appear in practical projects.

In Gruen [7] an example of two types of systematic errors, which are often met in practical projects, is given. In Gruen [9] the previous concept has been extended to include gross errors (blunders) as well, and has also been supported by examples.

Appendix B
Trace Check of Covariance Matrix

This procedure represents a computationally efficient technique to compute the effect of an individual additional parameter (or a group) on the trace of the covariance matrix of the other parameters of a bundle system, or a subset therefrom. It is based on the matrix identities

$$\tilde{N} = N + UWV\,, \tag{B1}$$

$$\tilde{N}^{-1} = N^{-1} - N^{-1}U\left(W^{-1} + VN^{-1}U\right)^{-1}VN^{-1}\,. \tag{B2}$$

Assume that a self-calibrating bundle system with a full set of additional parameters generates the normal equation matrix N and the related weight coefficient matrix Q_{xx}.

The deletion of one or more additional parameters leads to the normal equation matrix $\tilde{N}$ and to $\tilde{Q}_{xx}$ respectively. In order to delete the additional parameter of column (i) of N, a sufficiently large number is to be added to the diagonal element n_{ii} of N. In the context of (B1), the addition of a large number w_{ii} may be represented as

$$\tilde{N} = N + u_i w_{ii} v_i^T \tag{B3}$$

with

$$u_i^T = v_i^T = 0, \ldots, 1, \ldots, 0 ;$$
$$(1, \ldots, (i), \ldots, u) .$$

Thus w_{ii} may be interpreted as a weight of an a priori observed additional parameter. If the observation is assumed to be 0, a large weight causes the parameter (i) to be forced to this observed value 0. For $\tilde{N}^{-1}$ we obtain

$$\tilde{N}^{-1} = N^{-1} - N^{-1} u_i \left(\frac{1}{w_{ii}} + v_i^T N^{-1} u_i \right)^{-1} v_i^T N^{-1}, \tag{B4a}$$

or

$$\tilde{Q}_{xx} = Q_{xx} - Q_{xx} S Q_{xx} . \tag{B4b}$$

The matrix S given by

$$S = u_i \left(\frac{1}{w_{ii}} + v_i^T Q_{xx} u_i \right)^{-1} v_i^T \tag{B5a}$$

has the particular simple structure

$$S = \begin{bmatrix} 0 & . & . & . & 0 \\ . & . & & & . \\ . & . & s_{ii} & . & . \\ . & & & . & . \\ 0 & . & . & . & 0 \end{bmatrix} \tag{B5b}$$

with

$$s_{ii} = \frac{1}{1/w_{ii} + q_{ii}} . \tag{B6a}$$

q_{ii} is the ith diagonal element of Q_{xx} and for $w_{ii} \to \infty$,

$$s_{ii} = \frac{1}{q_{ii}} . \tag{B6b}$$

Applying the trace operator tr to (B4b), we obtain

$$\text{tr} \left(\tilde{Q}_{xx} \right) = \text{tr} \left(Q_{xx} \right) - \text{tr} \left(Q_{xx} S Q_{xx} \right) . \tag{B7}$$

With (B5b) and (B6b), the trace correction term $\Delta \text{tr} = \text{tr} \left(Q_{xx} S Q_{xx} \right)$ takes the particular simple form

$$\Delta \text{tr} = \text{tr}\,(\boldsymbol{Q}_{xx}\boldsymbol{S}\boldsymbol{Q}_{xx}) = \frac{1}{q_{ii}}\sum_{j=1}^{u}q_{ij}^2\,, \tag{B8}$$

where u is the number of system parameters in $\boldsymbol{N}$.

Hence, in order to check the influence of one additional parameter (i) on the trace of the covariance matrix, we need to compute only the ith row/column of $\boldsymbol{N}^{-1}$. If $\boldsymbol{N}$ is once factorized, the computational effort for that is no more than $O(A, M) = u^2 - 2u + 1$ (where $O(A, M)$ represents the number of additions and multiplications).

An equivalent expression to (B8) for the deletion of a set of addition parameters can readily be derived along the same lines.

Introducing the correlation coefficient ρ_{ij}, where

$$\rho_{ij}^2 = \frac{q_{ij}^2}{q_{ii}\,q_{jj}}\,, \tag{B9}$$

(B8) becomes

$$\Delta \text{tr} = \text{tr}\,(\boldsymbol{Q}_{xx}\boldsymbol{S}\boldsymbol{Q}_{xx}) = \sum_{j=1}^{u}\rho_{ij}^2 q_{jj}\,. \tag{B10}$$

Instead of checking the complete trace, it might be even more conclusive only to check the subtrace that is related to the object point coordinates. With u_i as the number of object point coordinates and $\boldsymbol{Q}_{xx}^{ui}$ the weight coefficient submatrix for object point coordinates, we would obtain

$$\Delta \text{tr} = \text{tr}\,\left(\boldsymbol{Q}_{xx}^{ui}\boldsymbol{S}\boldsymbol{Q}_{xx}^{ui}\right) = \frac{1}{q_{ii}}\sum_{j=1}^{u_1}q_{ij}^2 = \sum_{j=1}^{u_1}\rho_{ij}^2 q_{jj}\,; \qquad i \neq j \tag{B11}$$

Hence we get the alteration of the mean variance $\Delta\sigma_M^2$ of a network's point coordinates caused by one particular additional parameter (i)

$$\Delta\sigma_M^2 = \sigma_0^2\frac{1}{u_i}\sum_{j=1}^{u_i}\rho_{ij}^2 q_{jj}\,. \tag{B12}$$

Equation (B12) can be interpreted as a modified weighted mean, whereby the squared correlation coefficients ρ_{ij}^2 serve as "weighting factors" for the weight coefficients q_{jj} of the object point coordinates.

Individual variances may be checked by

$$\Delta q_{jj} = q_{jj} - \tilde{q}_{jj} = \rho_{ij}^2 q_{jj}\,, \tag{B13}$$

$$\tilde{\sigma}_{jj}^2 = \sigma_0^2\left(1 - \rho_{ij}^2\right)q_{jj}\,,$$
$$\Delta\sigma_{jj}^2 = \sigma_0^2\rho_{ij}^2 q_{jj}\,. \tag{B14}$$

Appendix C
Results of Computational Versions for the Determinability of Additional Parameters

In Tables C1–C7 the following symbols are used:

V	Version number
F	Number of frames
C	Configuration of images for given number of frames
P	Number of target planes
Co	Number of control points
Ch	Number of check points
$\hat{\sigma}_0$	Standard deviation of unit weight a posteriori
$\bar{\sigma}_X, \bar{\sigma}_Y, \bar{\sigma}_Z$	Mean standard deviations of check point coordinates
$\sigma_{X_0}, \sigma_{Y_0}, \sigma_{Z_0}$	Standard deviations of perspective center coordinates

Table C1. One frame.

V	F	P	Co	Ch	AP	$\hat{\sigma}_0$ [μm]	σ_{X_0} [mm]	σ_{Y_0} [mm]	σ_{Z_0} [mm]	Max. Correlation in % AP-OP	AP-EO	AP-AP
11030	1	1	3	0	0	n.a.	2.75	2.59	0.57			
11040	1	1	4	0	0	2.21	1.94	1.83	0.40			
					3	2.03	33.80	33.77	36.65	0.0	100.0	0.1
11050	1	1	5	0	0	17.5	1.65	1.73	0.40			
					3	17.5	32.26	32.24	36.48	0.0	100.0	0.3
					5	24.1	36.30	36.31	36.44	0.0	100.0	45.9
11090		1	9	0	0	21.7	1.36	1.55	0.32			
					3	21.6	30.94	30.92	36.16	0.0	100.0	1.6
					9	2.02	35.54	35.67	35.44	0.0	99.5	52.5
12050	1	2	5	0	0	12.8	0.64	0.72	0.40			
					3	1.75	1.03	6.25	35.96	0.0	100.0	92.5
12080	1	2	8	0	0	31.8	0.43	0.50	0.24			
					3	1.99	0.53	0.50	1.04	0.0	97.7	2.2
					6	0.87	0.55	0.70	7.39	0.0	99.6	99.8
					9	0.93	2.22	2.47	7.65	0.0	100.0	99.9

Table C2. Two frames, configuration 1. Note: Versions marked with "1" have been computed using standard deviations of 1.0 mm for $\Delta x_H, \Delta y_H, \Delta c$.

V	F C P Co Ch AP	$\hat{\sigma}_0$ [μm]	σ_{X_0} [mm]	σ_{Y_0} [mm]	σ_{Z_0} [mm]	Max. Correlation in % AP-OP	AP-EO	AP-AP	Note
211000 2 1 1 min 21	0	9.36	0.942	1.475	3.747				
	3	9.36	1.320	2.519	3.749	87.6	99.4	0.0	
	3	9.15	49.96	48.87	2.59	95.5	93.1	26.7	1
211040 2 1 1 4 20	0	9.67	0.648	0.618	2.748				
	3	8.99	0.697	0.637	2.809	54.8	100.0	3.1	
	3	8.74	0.478	0.432	1.898	95.1	100.0	17.9	1
	5	8.97	0.715	0.653	2.845	53.9	100.0	26.2	
211090 2 1 1 9 15	0	10.2	0.569	0.568	2.484				
	3	9.98	0.593	0.574	2.514	38.8	100.0	5.9	
	3	9.88	0.293	0.279	0.744	88.4	100.0	22.9	1
	9	6.11	0.626	0.633	2.770	45.6	94.1	92.4	
212000 2 1 2 min 33	0	8.66	1.102	1.798	3.219				
	3	8.62	1.835	4.269	4.928	97.6	99.6	8.3	
	3	8.60	12.853	33.032	26.929	99.8	99.7	84.1	1
212040 2 1 2 4 32	0	9.06	0.722	0.704	2.698				
	9	6.21	5.358	5.091	7.242	98.7	99.5	81.8	
212050 2 1 2 5 31	0	10.8	0.555	0.552	2.281				
	3	9.19	0.756	0.804	2.715	95.2	99.9	71.0	
	9	6.29	0.744	0.777	3.173	64.5	90.6	83.6	
212080 2 1 2 8 28	0	13.6	0.529	0.519	2.295				
	9	6.28	0.638	0.651	2.977	60.6	98.2	89.6	

Table C3. Two frames, configuration 2.

V	F C P Co Ch AP	$\hat{\sigma}_0$ [μm]	σ_{X_0} [mm]	σ_{Y_0} [mm]	σ_{Z_0} [mm]	Max. Correlation in % AP-OP	AP-EO	AP-AP
221000 2 2 1 min 21	0	7.68	0.693	0.612	1.092			
	3	7.68	5.542	6.862	1.094	99.4	99.9	0.0
221040 2 2 1 4 20	0	7.9	0.424	0.401	0.813			
	3	1.24	0.461	0.432	0.851	44.2	95.7	11.5
	9	0.99	3.243	3.025	1.109	97.0	99.8	
222000 2 2 2 min 33	0	12.5	0.694	0.700	0.912	99.8		98.4
	3	12.5	4.963	6.614	1.689		100.0	
	4	12.7	69.531	11.263	33.075	100.0	98.8	30.2
222040 2 2 2 4 32	0	12.2	0.445	0.465	0.738			
	3	11.6	0.469	0.4861	0.746	49.7	96.4	88.5
	9	1.05	2.03	1.510	2.660		99.7	
222050 2 2 2 5 31	0	12.4	0.406	0.267	0.649	99.3		6.8
	9	1.06	0.546	0.505	0.749		96.8	95.3
222080 2 2 2 8 28	0	15.8	0.376	0.343	0.632	76.3		99.5
	9	1.4	0.473	0.412	0.688	63.3	98.5	95.2

Table C4. Three frames.

V	F	P	Co	Ch	AP	$\hat{\sigma}_0$ [μm]	σ_{X_0} [mm]	σ_{Y_0} [mm]	σ_{Z_0} [mm]	Max. Correlation in % AP-OP	AP-EO	AP-AP
31000	3	1	min	21	0	3.09	0.496	0.536	0.924			
					3	3.08	3.376	6.306	1.017	99.5	99.9	80.8
					4	3.10	15.911	6.333	1.019	99.6	99.8	98.3
					9	1.26	16.286	8.144	1.035	99.1	99.7	97.8
31040	3	1	4	20	0	3.24	0.332	0.328	0.773			
					9	1.30	1.093	0.829	0.803	94.6	99.6	98.8
32000	3	2	min	33	0	11.5	0.514	0.644	0.823			
					3	9.12	0.860	1.713	0.880	97.7	99.4	42.3
					4	9.15	11.324	2.014	3.912	99.4	99.9	99.6
					9	1.22	12.308	2.868	4.288	99.9	99.8	99.5
32040	3	2	4	32	0	11.3	0.355	0.381	0.680			
					3	9.42	0.360	0.391	0.682	38.8	99.1	6.80
					9	1.29	0.794	0.466	0.758	94.4	99.5	99.1

Table C5. Four frames, configuration 1.

V	F	C	P	Co	Ch	AP	$\hat{\sigma}_0$ [μm]	σ_{X_0} [mm]	σ_{Y_0} [mm]	σ_{Z_0} [mm]	Max. Correlation in % AP-OP	AP-EO	AP-AP
411000	4	1	1	min	21	0	3.36	0.437	0.486	0.879			
						3	3.29	1.104	2.038	0.901	98.4	99.6	47.5
						4	3.31	5.210	2.217	0.960	99.4	99.6	96.3
						9	1.64	10.466	9.644	0.944	99.1	99.7	98.9
411040	4	1	1	4	20	0	3.39	0.294	0.289	0.726			
						3	3.36	0.309	0.298	0.756	35.1	98.6	8.2
						9	1.67	0.708	0.701	0.740	88.8	99.5	98.7
412000	4	1	2	min	33	0	9.71	0.454	0.591	0.767			
						3	8.89	0.571	0.914	0.800	89.2	97.2	13.1
						9	1.57	1.234	1.150	0.918	90.7	96.8	84.2
412040	4	1	2	4	32	0	9.61	0.320	0.335	0.633			
						9	1.60	0.446	0.410	0.653	78.0	97.6	96.0

Table C6. Four frames, configuration 2.

V	F	C	P	Co	Ch	AP	$\hat{\sigma}_0$ [μm]	σ_{X_0} [mm]	σ_{Y_0} [mm]	σ_{Z_0} [mm]	Max. Correlation in % AP-OP	AP-EO	AP-AP	Note
811000	8	1	1	min	21	0	3.18	0.348	0.370	0.652				
						9	1.52	0.475	0.486	0.658	77.7	87.2	83.9	1
						9	1.57	0.503	0.502	0.654	77.5	90.9	89.7	
811040	8	1	1	4	20	0	3.16	0.225	0.223	0.534				
						9	1.52	0.340	0.333	0.541	78.5	93.1	88.5	1
						9	1.57	0.355	0.347	0.538	79.1	95.0	92.3	
812000	8	1	2	min	33	0	7.94	0.356	0.448	0.579				
						9	1.49	0.375	0.464	0.581	33.7	87.5	90.8	1
						9	1.52	0.382	0.463	0.575	33.9	89.9	92.1	
812040	8	1	2	4	32	0	7.92	0.242	0.262	0.476				
						9	1.49	0.262	0.284	0.481	35.2	92.2	90.4	1
						9	1.52	0.265	0.287	0.478	35.0	93.5	93.4	

Table C7. Eight frames, configuration 1. Note: Versions marked with "1" have been computed using standard deviations of 0.01 mm for $\Delta x_H, \Delta y_H, \Delta c$.

V	F	C	P	Co	Ch	AP	$\hat{\sigma}_0$ [μm]	σ_{X_0} [mm]	σ_{Y_0} [mm]	σ_{Z_0} [mm]	Max. Correlation in % AP-OP	AP-EO	AP-AP
421000	4	2	1	min	21	0	2.6	0.490	0.433	0.768			
						3	1.82	3.329	0.554	0.778	99.2	99.9	58.7
						5	1.68	3.543	0.564	0.770	99.4	99.9	64.0
						9	1.38	3.768	1.164	0.803	99.0	99.1	90.9
421040	4	2	1	4	20	0	2.59	0.300	0.283	0.573			
						3	2.09	0.307	0.299	0.596	42.1	94.6	7.2
						9	1.37	0.756	0.723	0.632	90.2	96.3	94.2
422000	4	2	2	min	33	0	6.68	0.500	0.663				
						9	1.36	2.225	0.968	0.941	98.7	99.6	90.3
422040	4	2	2	4	32	0	6.99	0.323	0.347	0.538			
						3	6.53	0.333	0.347	0.544	47.6	96.3	7.1
						9	1.39	0.365	0.557	46.7	96.2	94.3	

References

1. A. Gruen. Towards real-time photogrammetry. Photogrammetria **42**, 209–244, 1988
2. H. Beyer. Some aspects of the geometric calibration of CCD cameras. *Proceedings ISPRS Intercommission Conference on Fast Processing of Photogrammetric Data*, pp. 68–81. Interlaken, Switzerland, June 1987
3. H. Beyer. Geometric and radiometric analysis of a CCD camera based photogrammetric close-range system. Dissertation, No. 9701, ETH, Zürich 1992
4. A. Gruen, H. Beyer: Real-time photogrammetry at the Digital Photogrammetric Station (DIPS) of ETH Zurich. Paper presented at the ISPRS Commission V Symposium, Real-Time Photogrammetry – A New Challenge, June 16–19, Ottawa, and in: The Canadian Surveyor **41**(2), 181–199, 1986
5. D.C. Brown. A solution to the general problem of multiple station analytical stereotriangulation. RCA Data Reduction Technical Report, No. 43, Patrick Air Force Base, Florida, 1958
6. D.C. Brown. Close-range camera calibration. Photogrammetric Engineering, **37**(8), 855–866, 1971
7. A. Gruen. Progress in photogrammetric point determination by compensation of systematic errors and detection of gross errors. Nachrichten aus dem Karten- und Vermessungswesen, Series II, **36**, 113–140, 1978
8. A. Gruen. Precision and reliability aspects in close-range photogrammetry. Phot. Journal of Finland **8**(2), 117–132, 1980
9. A. Gruen. Photogrammetric point positioning with bundle adjustment (in German). Institut für Geodäsie und Photogrammetrie, ETH Zürich, Mitteilungen Nr.40. (1986)
10. E. Kilpelä. Compensation of systematic errors of image and model coordinates (Report of ISPRS WG III/3). International Archive of Photogrammetry, **23**, B9, Hamburg 1980
11. D.C. Brown. The bundle adjustment – progress and prospects. International Archive of Photogrammetry **21** B3, Helsinki 1976
12. A. Gruen. Experiences with self-calibrating bundle adjustment. Presented Paper to the Convention of ACSM/ASP, Washington, D.C., March 1978
13. A. Gruen. Data processing methods for amateur photographs. Photogrammetric Record **11**(65), 567–579, 1985

8 Self-Calibration of a Stereo Rig from Unknown Camera Motions and Point Correspondences

Quang-Tuan Luong and Olivier D. Faugeras

Summary

The problem of calibrating a stereo rig is extremely important for practical applications. Existing work is based on the use of a calibration pattern whose 3D model is a priori known. We show theoretically and with experiments on real images, how it is possible to completely calibrate a stereo rig, that is to determine each camera's intrinsic parameters and the relative displacement between the two or three cameras, using only point matches obtained during unknown motions, without any a priori knowledge of the scenes.

The first part of the chapter is devoted to the computation of the intrinsic parameters of the cameras by a method based upon the estimation of the so-called fundamental matrix associated with camera displacement. Three different displacements are sufficient to solve the Kruppa equations which yield these parameters.

The second part of the chapter is devoted to the computation of the extrinsic parameters. We first explain how to recover the unknown motions previously used, once we have an estimate of the intrinsic parameters and the fundamental matrices. The computation is quite robust to the inaccuracy of the determination of the camera parameters. We then present the equations which allow us, from two displacements of the stereo rig, for which the camera motions are computed independently, to compute the relative displacement between the cameras. This technique allows us to compute the relative displacement between two or three cameras and complete the full calibration of the rig.

8.1 Introduction

8.1.1 The Stereo Calibration Problem

Camera calibration is an important task in computer vision. The purpose of camera calibration is to establish the relationship between the 3D world

Springer Series in Information Sciences, Vol. 34
Calibration and Orientation of Cameras in Computer Vision
Eds.: Gruen, Huang © Springer-Verlag Berlin Heidelberg 2001

coordinates and their corresponding 2D image coordinates. Once this relationship is established, 3D information can be inferred from 2D information, and vice versa. Thus camera calibration is a prerequisite for any application where this relation between 2D images and the 3D world is needed. In the case of a stereo rig, calibration is needed for at least two reasons. First to establish the epipolar geometry of the system and cut down the complexity of the stereo correspondence process. Second, to reconstruct the 3-D data after matching between the retinas has been completed.

The model which we consider is the most widely used. It is the pinhole model. The basic assumption behind this model is that *the relationship between the world coordinates and the pixel coordinates is linear projective.* Thus no camera distortion is considered which allows us to use the powerful tools of projective geometry, which is emerging as an attractive framework for computer vision [1]. In this chapter, we assume that the reader is familiar with the elementary projective geometry described in [2] for example. The equation of the model for one camera is:

$$\begin{bmatrix} su \\ sv \\ s \end{bmatrix} = \boldsymbol{A}_1 \begin{bmatrix} 1\ 0\ 0\ 0 \\ 0\ 1\ 0\ 0 \\ 0\ 0\ 1\ 0 \end{bmatrix} \boldsymbol{D}_1 \begin{bmatrix} X \\ Y \\ Z \\ 1 \end{bmatrix} = \boldsymbol{P}_1 \begin{bmatrix} X \\ Y \\ Z \\ 1 \end{bmatrix} \qquad (8.1)$$

where u, v are retinal coordinates, X, Y, Z, world coordinates, $\boldsymbol{A}_1$ a 3×3 transformation matrix accounting for camera sampling and optical characteristics, $\boldsymbol{D}_1$ a 4×4 displacement matrix accounting for camera position and orientation: $\boldsymbol{D}_1$ is the displacement from the world coordinate system (identified to the object that has been used for the calibration) to the camera coordinate system. The 3×4 matrix $\boldsymbol{P}_1$ is the perspective projection matrix, which relates 3D world coordinates and 2D retinal coordinates. $\boldsymbol{A}_1$ depends on a variable number of parameters, depending on the sophistication of the camera model: these parameters are called intrinsic. We will consider here a five-parameter model, represented in Fig. 8.1, which we explain later. $\boldsymbol{D}_1$ depends on six parameters, called extrinsic: three defining a rotation, three a translation, and has the form:

$$\boldsymbol{D}_1 = \begin{pmatrix} \boldsymbol{R}_1\ \boldsymbol{t}_1 \\ \boldsymbol{0}_3^{\mathrm{T}}\ 1 \end{pmatrix}. \qquad (8.2)$$

Using a second camera introduces another set of intrinsic parameters through matrix $\boldsymbol{A}_2$, and another set of extrinsic parameters through matrix $\boldsymbol{D}_2$. However, in general the choice of a particular world coordinate system versus another one is arbitrary, the significant thing being the relative position and orientation of the two cameras; thus we can consider that there are only six significant extrinsic parameters for a binocular stereo rig. They are given by the 4×4 displacement matrix $\boldsymbol{D}$, representing the displacement from the first camera to the second camera, which can be easily computed

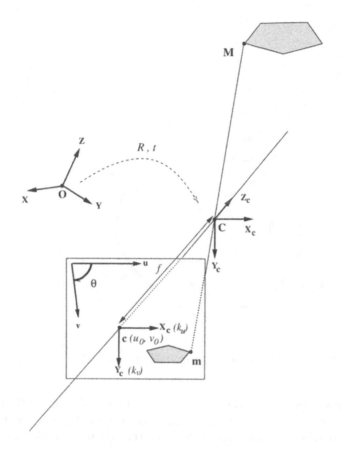

Fig. 8.1. The general projective camera model.

from D_1 and D_2. If we don't consider a coordinate system linked to an object of the world, we must express this displacement in a coordinate system linked to one of the two cameras, as shown Fig. 8.2. Then, fully calibrating a binocular stereo rig means to compute:

- the five intrinsic parameters of the first camera, represented by A_1
- the five intrinsic parameters of the second camera, represented by A_2
- the six extrinsic parameters represented by D, displacement from the first to the second camera, in the coordinate system of the first camera.

In the case of a trinocular stereo rig, we have of course five more intrinsic parameters represented by A_3 and 6 more extrinsic parameters represented D' to determine.

8.1.2 What do we Mean by Self-Calibration

In the usual method of calibration [3,4] a special object (calibration grid) is put in the field of view of the cameras. It is assumed that we have a 3D model

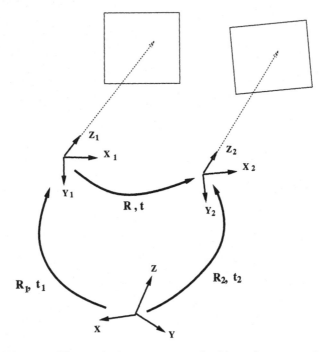

Fig. 8.2. The extrinsic parameters of a binocular stereo rig.

of this object, that is we know the 3D coordinates of some of its reference points, in a coordinate system attached to the object. Usually a regular pattern is used, which is painted in such a way that retinal coordinates of points of interest (for example, corners) can be measured with great accuracy. Using a large number of points, each one yielding an equation of the form (8.1) the perspective transformation matrix M can then be estimated independently for each camera. This method has been widely used and yields a very good accuracy in the determination of the camera parameters, provided the reference grid is carefully set. The drawback of this method is that in many applications a calibration grid is not available. Another drawback is that it is not possible to calibrate on-line, when the camera is already involved in a visual task. However, even when the camera performs a task, the intrinsic and extrinsic parameters can change, intentionally (for example adjustment of the focal length or of the vergence), or not (mechanical or thermal variations).

The goal of our work is to elaborate a general calibration method that can be carried out using the same images as those that are used to perform the visual task. No a priori knowledge of the scenes is needed. The only requirements are that the stereo rig undergoes a series of displacements in a rigid scene and that we are capable of establishing correspondences between points in pairs of images taken by the same camera at different positions. By this we mean identifying points in retinas that are images of the same point in the scene. Our method, which we call self-calibration:

- is automatic,
- does not require any model of the observed objects,
- does not require any knowledge about the camera motion,
- does not require any a priori knowledge about the camera parameters.

As we do not use any metric information, we can only compute the extrinsic parameters only up to a scale factor: our method recovers only five extrinsic parameters, the three defining the rotation, and two defining the direction of the translation. In the case of a binocular stereo rig, this is not a problem, as this information is sufficient to obtain the epipolar geometry of the stereo system, and to perform 3D reconstruction up to a scale factor. Its determination could be obtained very simply, by just showing to one camera an object of known length. When self-calibrating a trinocular stereo rig, in addition to the $10 = 5 \times 2$ extrinsic parameters, the ratio of the amplitudes of the translations are needed. We also propose a method to obtain this information. Thus we are able to recover metric information using well-established algorithms based on strong calibration, *up to a scale factor*. To provide metric information in a traditional form, the self-calibration algorithm outputs two or three projection matrices, the first one being expressed in the coordinate frame of the first camera.

8.1.3 An Outline of our Autonomous Approach

We first need to establish point correspondences between pairs of images taken by the same camera. This stage can be done in two steps, which will not be described in detail in this chapter. The reader is referred to the references.

Extraction of points of interest: Since we do not use a model of a particular object, we must use characteristic variations of intensity that are general. The features that we found to be the most useful are corners, as some vertexes (triple junctions) result from occlusion. There are three main approaches to the problem of corner detection:

- The first one is to extract features such as edges chains [5], [6] or polygonal approximations [7], and then to search for corners using these data. Apart from the computationnal cost, these methods suffer from the possible difference of the intermediate features extracted in the various images.
- The second one [8], [9] is to first apply a differential operator measuring gradients and curvatures of the image intensity surface, and then to select points that are corners by a second operator that is often a thresholding scheme. It is a global and quite efficient approach, however it has been shown [10] that the most notorious algorithms of this family yield a precision in the positionning of only a few pixels.
- The third one [11–13] is to use an explicit model of the local image structure in the neighborhood of the target corner, and to search the numerical parameters of such a model by a nonlinear minimization. The advantage

of this approach is the subpixel accuracy in localization, which is necessary for self-calibration, as will be seen later. The main drawback is the high computationnal cost, and the difficulty of performing the method in a purely automatic manner.

Correspondences: Once the corners have been obtained independently in each image, the correspondences can be obtained using standard correlation techniques [14–16] in the general discrete case, or by a tracking technique, such as the one presented in [17], in the case when a long sequence of images is available. Note that an initial estimate of the epipolar geometry, obtained with at least seven point matches, can be used to further refine the correspondence process.

The goal of this chapter is to present an approach which starts from the correspondences and leads to a complete metric calibration of a stereo rig. The applicability of the whole approach is illustrated by examples of tridimensionnal reconstruction from triplets of real, uncalibrated images.

From point correspondences to fundamental matrices: The first step is the computation of the so-called fundamental matrix for each camera motion. The knowledge of this matrix is equivalent to that of the epipolar transformation, which contains all the geometric information which is possible to obtain from two uncalibrated images. We present a nonlinear approach which is far more accurate and robust than the linear approaches found in the literature.

From fundamental matrices to intrinsic parameters: The second step is the determination of the intrinsic parameters of the cameras. Each fundamental matrix yields two Kruppa equations. When we have done enough movements, we can solve them for the coefficients of the equation of the image of the absolute conic (to be explained later), from which it is easy to obtain the intrinsic parameters of the camera.

From fundamental matrices and intrinsic parameters to cameras motions: The third step is the recovery of the unknown motions previously used, once we know the intrinsic parameters. These motions are computed in a coordinate system associated with the cameras, using two different methods that we compare, one based on a minimization procedure on a criterion very similar to the one we use for the computation of the fundamental matrix, the other using a closed-form solution involving directly the fundamental matrices previously found.

From cameras motions to stereo extrinsic parameters: The fourth step is an approach which allows us, from two displacements of the stereo rig during which motions are computed independently in each camera's coordinate system, to compute the relative displacement between the two cameras. The method is to solve a set of matrix equations using the particular structure of the motion resulting from the rigidity of the stereo rig. A variant of the method allows us to compute the relative displacements between three cameras.

8.2   Computing the Fundamental Matrix

Almost all the point-based algorithms which start from multiple, uncalibrated images [18], [19], [20], [21] require, as the basic information, the fundamental matrix, which is the only alternative to projection matrices, in order to relate two views of the same scene. It is also the case for the self-calibration algorithm. Eight points are needed to obtain a unique solution.

8.2.1   The Fundamental Matrix and the Epipolar Transformation

The Epipolar Geometry and the Fundamental Matrix. The epipolar geometry is the basic constraint which arises from the existence of two viewpoints. Let a camera take two images by linear projection from two different locations, as shown in Fig. 8.3. Let C be the optical center of the camera when the first image is obtained, and let C' be the optical center for the second image. The line $\langle C, C' \rangle$ projects to a point e in the first image $\mathcal{R}$, and to a point e' in the second image $\mathcal{R}'$. The points e, e' are the epipoles. The lines through e in the first image and the lines through e' in the second image are the epipolar lines. The epipolar constraint is well-known in stereovision: for each point m in the first retina, its corresponding point m' lies on its epipolar line l'_m.

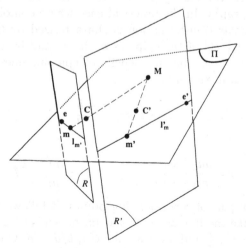

Fig. 8.3. The epipolar geometry.

Let us now use retinal coordinates. The relationship between the projective representation of a point m and its corresponding epipolar line l'_m is projective linear, because the relations between m and $\langle C, M \rangle$, and m and $\langle C, M \rangle$ and its projection l'_m are both projective linear. We call the 3×3 matrix F which describes this correspondence the *fundamental matrix*. Let us now express the epipolar constraint using the fundamental matrix, in the case of uncalibrated cameras. For a given point m in the first image, the projective representation l'_m of its the epipolar line in the second image is given by

$$l'_m = Fm .$$

Since the point m' corresponding to m belongs to the line l'_m by definition, it follows that

$$m'^{\mathrm{T}} Fm = 0 . \tag{8.3}$$

The importance of the fundamental matrix has been overlooked in the literature, since almost all the work on motion has been done under the assumption that the intrinsic parameters were known. In that case, the fundamental matrix reduces to an essential matrix [22]. But if one wants to proceed only from image measurements, the fundamental matrix is the key concept, since it contains all the geometrical information relating two different images.

Relation with the Epipolar Transformation. The epipolar transformation is a homography between the epipolar lines in the first image and the epipolar lines in the second image, defined as follows. Let Π be any plane containing $\langle C, C' \rangle$. Then Π projects to an epipolar line l in the first image and to an epipolar line l' in the second image. The correspondences $\Pi \leftrightarrow l$ and $\Pi \leftrightarrow l'$ are homographies between the two pencils of epipolar lines and the pencil of planes containing $\langle C, C' \rangle$. It follows that the correspondence $l \leftrightarrow l'$ is a homography. In the practical case where epipoles are at a finite distance, the epipolar transformation is characterized by the *affine* coordinates $[e_1, e_2]$ and $[e'_1, e'_2]$ of the epipoles e and e' and by the coefficients of the homography between the two pencils of epipolar lines, *each line being parameterized by its direction τ*:

$$\tau \mapsto \tau' = \frac{a\tau + b}{c\tau + d} \tag{8.4}$$

where

$$\tau = \frac{m_2 - e_2}{m_1 - e_1} \qquad \tau' = \frac{m'_2 - e'_2}{m'_1 - e'_1} \tag{8.5}$$

and $m \leftrightarrow m'$ is a pair of corresponding points. It follows that the epipolar transformation, like the fundamental matrix, depends on seven independent parameters, e_1, e_2, e'_1, e'_2 and for example $a/d, b/d, c/d$. On identifying (8.3) with the constraint on epipolar lines obtained by making the substitutions (8.5) in (8.4), expressions are obtained for the coefficients of F in terms of the parameters describing the epipoles and the homography. Conversely, (8.6), yield the epipolar transformation as a function of the fundamental matrix:

$$
\begin{aligned}
a &= F_{12} \\
b &= F_{11} \\
c &= -F_{22} \\
d &= -F_{21} \\
e_1 &= \frac{F_{23}F_{12} - F_{22}F_{13}}{F_{22}F_{11} - F_{21}F_{12}} \\
e_2 &= \frac{F_{13}F_{21} - F_{11}F_{23}}{F_{22}F_{11} - F_{21}F_{12}} \\
e'_1 &= \frac{F_{32}F_{21} - F_{22}F_{31}}{F_{22}F_{11} - F_{21}F_{12}} \\
e'_2 &= \frac{F_{31}F_{12} - F_{11}F_{32}}{F_{22}F_{11} - F_{21}F_{12}} \ .
\end{aligned}
\tag{8.6}
$$

The determinant $ad - bc$ of the homography is $F_{22}F_{11} - F_{21}F_{12}$. In the case of finite epipoles, it is nonzero. The interpretation (8.6) is simple: the vectors e (resp. e') are in the nullspace of F (resp. F^{T}). This implies in particular that the rank of F is less than or equal to 2. In practice, it is equal to 2 because a rank of 1 would imply that all epipolar lines are the same. Therefore, we have $\det(F) = 0$. Writing τ' as a function of τ from the relation $m'_\infty F m_\infty = 0$ which arises from the correspondence of the points at infinity $m_\infty = (1, \tau, 0)^{\mathrm{T}}$ and $m'_\infty = (1, \tau', 0)^{\mathrm{T}}$, of corresponding lines, we obtain the homographic relation.

8.2.2   A Robust Method for the Determination of the Fundamental Matrix

A first method to estimate the fundamental matrix takes advantage of the fact that (8.3) is linear and homogeneous in the nine unknown coefficients of F. Thus if eight matches (m_i, m'_i) are given then in general F is determined up to a scale factor. In practice, many more than eight matches are given. A linear least squares method can then be used to solve for F:

$$
\min_{F} \sum_i (m_i'^{\mathrm{T}} F m_i)^2 \quad \text{subject to } \mathrm{Tr}(F^{\mathrm{T}} F) = 1 \ .
\tag{8.7}
$$

We denote this method by **LIN**. The advantage of the quadratic criterion is that it leads to a noniterative computation method, however, we have found that it is quite sensitive to noise, even with numerous data points. The two main reasons for this are:

- The constraint $\det(F) = 0$ is not satisfied, which causes inconsistencies of the epipolar geometry near the epipoles.

- The criterion is not normalized, which causes a bias in the localization of the epipoles.

Experiments show that this problem is reduced by using the following criterion (denoted as **DIST**) for minimization:

$$\min_F \{ d(m'^{\mathrm{T}}, Fm)^2 + d(m^{\mathrm{T}}, F^{\mathrm{T}}m')^2 \} \qquad (8.8)$$

where d is a distance in the image plane. The criterion has a better significance in terms of distances in measurement space, and it is normalized, which means that it is invariant by a change of scale factor of F. It is necessary to include both terms in the criterion to keep the symmetry of the two-camera system to avoid discrepancies in the epipolar geometry. In order to achieve this nonquadratic minimization successfully, it is important to take into account the following two constraints:

- The solution must be of rank two, as all fundamental matrices have this property. Rather than performing a constrained minimization with the cubic constraint $\det(F) = 0$, it is possible to use, almost without loss of generality, a parameterization which accounts directly for this property, for instance by writing that row of F as a linear combination of the other two.
- The matrix F is defined only up to a scale factor. In order to avoid the solution $F = 0$, one of the elements of the first two rows of F must be normalized by giving it a fixed finite value. However, as the minimization is nonquadratic, convergence results can differ depending on the element chosen.

This second method for computing the fundamental matrix is more complicated, as it involves nonquadratic minimizations. However, it yields more precise results. We use the quadratic method to obtain a starting point. For a far more detailed analysis of different methods to compute the fundamental matrix, see [23].

8.3   Computing the Intrinsic Parameters of the Cameras

Once a camera has performed at least three displacements it is possible to solve for all its intrinsic parameters, using the fundamental matrices. If we restrict ourselves to the realistic case of an orthogonal pixel grid, two displacements are sufficient.

8.3.1   The Principle of the Method

The Longuet-Higgins equation [22], applies when using normalized coordinates, and thus calibrated cameras. If the displacement between the two positions of the cameras is given by the rotation matrix R and the translation

t, and if m and m' are corresponding points, then the coplanarity constraint relating $C'm'$, t, and Cm is written as:

$$m' \cdot (t \times Rm) \equiv m'^{\mathrm{T}} Em = 0 \, . \tag{8.9}$$

The matrix E, which is the product of an orthogonal matrix and an antisymmetric matrix, is called an essential matrix. Because of the depth/speed ambiguity, E depends on five parameters only.

It can be seen that the two (8.9) and (8.3) are equivalent, and that we have the relation:

$$F = A^{-1\mathrm{T}} E A^{-1}$$

The essential matrix E depends only on five independent parameters, it is thus subject to two independent polynomial constraints in addition to the constraint $\det(E) = 0$. If F is known then it follows from $E = A^{\mathrm{T}} F A$ that the entries of A are subject to two independent polynomial constraints inherited from E. The self-calibration method consists in using these constraints to obtain the intrinsic parameters from the fundamental matrices. If we use the most general model for A, we have five intrinsic parameters to compute, thus three displacements are needed. If we use a simplified model in which we suppose that the pixel grid is orthogonal, which is almost always the case, we have only four intrinsic parameters to compute, and thus two displacements are sufficient.

8.3.2   Kruppa Equations
Arising from an Epipolar Transformation

The previously mentioned constraints express the fact that the actual motion of the camera is necessarily a rigid displacement. Several equivalent formulations are possible. The most interesting formulation is to use the Kruppa equations [24], first introduced in the field of computer vision by Faugeras and Maybank for the study of motion [25] , and then to develop a theory of self-calibration [26]. The Kruppa equations are obtained from a geometric interpretation of the rigidity constraints: the tangents to the image ω of the absolute conic[1] in two views taken by the same camera correspond under the epipolar transformation. The matrix of the dual conic of ω is the dual matrix (matrix of cofactors) $K = B^*$, where B is the matrix of ω:

$$K = \begin{pmatrix} -\delta_{23} & \delta_3 & \delta_2 \\ \delta_3 & -\delta_{13} & \delta_1 \\ \delta_2 & \delta_1 & -\delta_{12} \end{pmatrix} \tag{8.10}$$

The epipolar line $l = \langle e, y \rangle$ is tangent to ω iff:

$$(e \times y)^{\mathrm{T}} K (e \times y) = 0$$

---

[1] The absolute conic is an imaginary circle of radius $i = \sqrt{-1}$ in the plane at infinity.

by parameterizing the epipolar line l with the projective parameter τ such that $\boldsymbol{y} = (1, \tau, 0)^{\mathrm{T}}$, this equation can be written:

$$P_1(\tau) = k_{11} + 2k_{12}\tau + k_{22}\tau^2 = 0$$

where the coefficients k_{11}, k_{12}, k_{22} are defined by

$$k_{11} = -\delta_{13} - \delta_{12}e_2^2 - 2\delta_1 e_2$$
$$k_{12} = \delta_{12}e_1 e_2 - \delta_3 + \delta_2 e_2 + \delta_1 e_1$$
$$k_{22} = -\delta_{23} - \delta_{12}e_1^2 - 2\delta_2 e_1 \ . \tag{8.11}$$

Similarly, the epipolar line l' through $\boldsymbol{p}'$ corresponding to l is tangent to ω:

$$k'_{11}(c\tau + d)^2 + 2k'_{12}(c\tau + d)(a\tau + b) + k'_{22}(a\tau + b)^2 = 0$$

which can be written:

$$P_2(\tau) = k''_{11} + 2k''_{12}\tau + k''_{22}\tau^2 = 0$$

with:

$$k''_{11} = k'_{22}b^2 + k'_{11}d^2 + 2k'_{12}bd;$$
$$k''_{12} = 2k'_{12}ad + 2k'_{22}ab + 2k'_{11}cd + 2k'_{12}bc$$
$$k''_{22} = 2k'_{12}ac + k'_{22}a^2 + k'_{11}c^2$$

the coefficients k'_{11}, k'_{12}, k'_{22} being obtained from (8.11) by replacing the coordinates e_i of $\boldsymbol{e}$ with the coordinates e'_i of $\boldsymbol{e}'$.

The polynomials P_1 and P_2 must have the same roots, which yield two independent so-called Kruppa equations, for example:

$$k_{22}k''_{12} - k''_{22}k_{12} = 0 \tag{8.12}$$
$$k_{11}k''_{12} - k''_{11}k_{12} = 0 \ . \tag{8.13}$$

These equations are of degree two in the six parameters δ_i, δ_{ij}, which are defined up to a scale factor. We call them the Kruppa coefficients.

8.3.3 Kruppa Coefficients and Intrinsic Parameters

We now underline the relation between the absolute conic and the intrinsic parameters in order to write the Kruppa coefficients as a function of the intrinsic parameters. The most general matrix $\boldsymbol{A}$ can be written as:

$$\boldsymbol{A} = \begin{bmatrix} -fk_u & fk_u \cot\theta & u_0 \\ 0 & -\dfrac{fk_v}{\sin\theta} & v_0 \\ 0 & 0 & 1 \end{bmatrix} \tag{8.14}$$

where (see Fig. 8.1):

- k_u, k_v, are the horizontal and vertical scale factors whose inverse characterize the size of the pixel, in world coordinates units.
- u_0 and v_0 are the image center coordinates, resulting from the intersection between the optical axis and the image plane.
- f is the focal length.
- θ is the angle between the directions of retinal axes. This parameter is introduced to account for the fact that the pixel grid may not be exactly orthogonal. In practice it is very close to $\frac{\pi}{2}$.

As we cannot separate f from k_u and k_v, we let $\alpha_u = -f k_u$ and $\alpha_v = -f k_v$, so we obtain five intrinsic parameters. This is exactly the number of independent coefficients for the absolute conic. The fact that the form (8.14) is the most general comes from the two equivalent facts that there is a unique decomposition (8.1) of each nondegenerate projection matrix, and that each plane conic without real points can be written $\boldsymbol{B} = \boldsymbol{A}^{-1\mathrm{T}}\boldsymbol{A}^{-1}$. From this last relation it follows that:

$$\delta_1 = v_0$$

$$\delta_2 = u_0$$

$$\delta_3 = u_0 v_0 - \alpha_u \alpha_v \frac{\cot\theta}{\sin\theta}$$

$$\delta_{12} = -1$$

$$\delta_{23} = -u_0^2 - \frac{\alpha_u^2}{\sin^2\theta}$$

$$\delta_{13} = -v_0^2 - \frac{\alpha_v^2}{\sin^2\theta} \ . \tag{8.15}$$

8.3.4 Solving the Kruppa Equations

Two approaches have been presented elsewhere in order to solve the Kruppa equations. The first one [27] takes advantage of the fact that these equations are polynomial, of degree two in the Kruppa coefficients. Thus, having done three displacements, we can use semi-analytical methods to solve the resulting polynomial system of six equations in six homogeneous unknowns. This is done by a numerical continuation method [28]: the idea is to intersect the six sets of $32 = 2^5$ solutions obtained by solving five equations only. A set of algebraic constraints must be verified by the Kruppa coefficients to ensure the existence of a real solution. These constraints turn out to be equivalent to the fact that ω has no real points. They allow us to discard spurious solutions. The main advantage of this approach is that no initial guess is needed at all, and thus it can be used even with no a priori knowledge of the intrinsic parameters, which is important.

The second approach is to use iterative methods [29], which have some practical advantages:

- it is easy to use long sequences,

- uncertainty and a priori knowledge can be easily taken into account,
- they are computationnaly efficient and ensure the existence of a real solution.

The idea of these methods is to substitute the values of the Kruppa coefficients (8.15) and of the parameters of the epipolar transformation (8.6) into the Kruppa (8.13) in order to obtain measurement equations which relate directly the entries of the fundamental matrices to the intrinsic parameters. These equations can then be solved using either an extended Kalman filter technique, or a batch minimization technique. In this last case, the choice of the criterion is important. We have found that the following criterion gives good results:

$$\min_{\alpha_u, \alpha_v, u_0, v_0, \theta} \sum_i \left(\frac{k_{11}}{k''_{11}} - \frac{k_{12}}{k''_{12}} \right)^2 + \left(\frac{k_{12}}{k''_{22}} - \frac{k_{12}}{k''_{22}} \right)^2 + \left(\frac{k_{11}}{k''_{22}} - \frac{k_{11}}{k''_{22}} \right)^2. \quad (8.16)$$

For more details on solving the Kruppa equations, and numerous simulations the reader is referred to [30].

8.4   Computing the Motion of the Camera

We suppose now that we have obtained the intrinsic parameters A of a camera. Our next goal is to compute the three-dimensionnal motion from pairs of images. This computation can be done quite robustly even with imprecise camera parameters.

8.4.1   Two Approaches Based on the Computation of the Fundamental Matrix

The motion determination problem from point correspondences is very classical. See [31–34] for solutions similar to ours. We present two different solutions, both based on the computation of the fundamental matrix.

A Direct Factorization. We have seen that during the course of intrinsic parameters estimation, we had to compute the fundamental matrix F, from which the essential matrix is immediately obtained:

$$E = A^{\mathrm{T}} F A. \quad (8.17)$$

The problem of finding the rotation R and the translation t from E is classical [22], [35], [31].

As we have, by construction, found an F-matrix of rank two, the direction of translation is just obtained by solving $E^{\mathrm{T}} t = 0$.

To find the rotation, we use a method introduced in [31]: in the presence of noise, we minimize with respect to the rotation matrix R the criterion:

$$C = \sum_{i=1}^{3} \| E_i - R^\mathrm{T} T_i \|^2$$

where E_i and T_i are the three lines of the matrices E and T, respectively. Using q a quaternion representing R, some properties of this representation yield:

$$C = \sum_{i=1}^{3} | q \times E_i - T_i \times q |^2 \qquad (8.18)$$

where $\times$ denotes the quaternion product. It follows from the definition of the quaternion product that $q \times E_i - T_i \times q$ is a linear function of the four coordinates of q. Therefore, there exists a 4×4 matrix N_i such that:

$$| q \times E_i - T_i \times q | = N_i q \quad \text{with} \quad N_i = \begin{pmatrix} 0 & (E_i - T_i)^\mathrm{T} \\ T_i - E_i & \tilde{E}_i + \tilde{T}_i \end{pmatrix}. \qquad (8.19)$$

Therefore, the problem reduces to a linear least-squares problem:

$$\min_{q} \sum_{i=0}^{3} q N_i N_i^\mathrm{T} q^\mathrm{T} \quad \text{subject to the constraint:} \quad \|q\|^2 = 1$$

which is a classical minimization problem, whose solution is the eigenvector associated with the smallest eigenvalue of $N = \sum_{i=1}^{3} N_i N_i^\mathrm{T}$. It can be noted that this solution is entirely equivalent to the well-known method of Tsai and Huang [35] , which has been recently proved to be optimal by Hartley [36]. We denote this algorithm by **FACTOR**.

An Iterative Solution. An alternative method is to use directly the criterion that has been used to determine the fundamental matrix. We denote by **MIN-LIN**[2] the minimization of the error criterion (8.7) and by **MIN-DIST** the minimization of the error criterion (8.8). The knowledge of the intrinsic parameters allows us to minimize these criteria with respect to five motion parameters: we parameterize T by t_1/t_3, t_2/t_3 and R by the three-dimensional vector r whose direction is that of the axis of rotation and whose norm is equal to the rotation angle. We use, as a starting point for this nonlinear minimization, the result obtained by **FACTOR**.

---

[2] Although it is not a linear method, but a nonlinear method based on the same error measure as the linear criterion for the computation of the fundamental matrix.

8.4.2   An Experimental Comparison

The Case of Exact Intrinsic Parameters. In the first comparative study, we suppose that the *exact* intrinsic parameters are known. The graphs have been obtained using 200 different displacements, and show the average relative error on the rotational and translational components. As the nonlinear methods need a starting point whose choice is important, we have considered three possibilities:

1. the exact motion, to test the precision of the minimum (Figs. 8.4 and 8.5).
2. the motion obtained by **FACTOR**, which is the realistic initialization (Figs. 8.6 and 8.7).
3. an arbitrary motion: $r = (\frac{1}{2}, \frac{1}{2}, \frac{1}{2})^{\mathrm{T}}$, $t = (0,0,1)^{\mathrm{T}}$, to test the convergence properties (Figs. 8.6 and 8.7).

The conclusions of the simulations are:

- The computation is more stable than the fundamental matrix computation. Motion computation is a less difficult problem.
- The rotational part is determined more precisely than the translational part.
- The iterative method based on **MIN-DIST** is the most precise, but it is the most sensitive to the choice of the starting point.
- The results obtained by **MIN-DIST** and by **FACTOR** in the realistic case are very close.

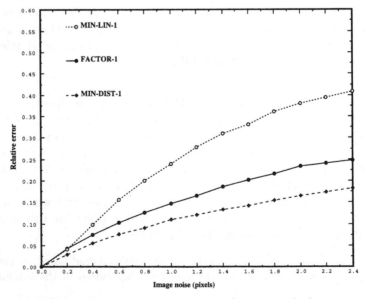

Fig. 8.4. Relative error on the rotation, initialization with the exact point.

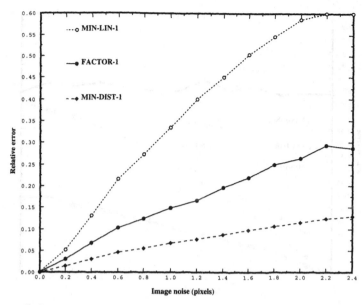

Fig. 8.5. Relative error on the translation, initialization with the exact point.

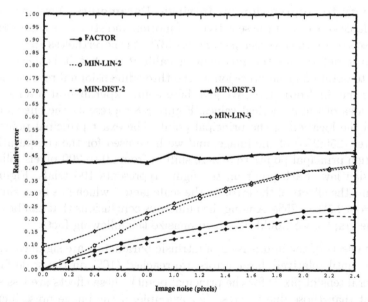

Fig. 8.6. Relative error on the rotation, initialization with two different values.

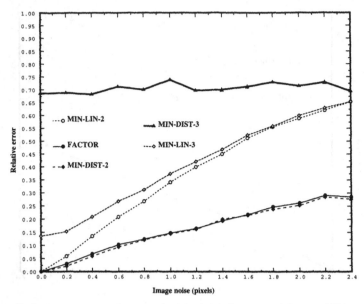

Fig. 8.7. Relative error on the translation, initialization with two different values.

Sensitivity to Imprecision on Intrinsic Parameters. Very few results are available concerning the sensitivity of motion and structure computations to imprecision on the intrinsic parameters [37]. It is nevertheless an important issue, as it determines the precision of calibration that it is necessary to achieve to obtain a given precision on the three-dimensionnal reconstruction, which is the final objective. We give here some experimental results which give an idea of the numerical values. Figure 8.8 represents the effects of the error on the location of the principal point. The exact principal point is at the center (255,255) of the image, and we have used for the computation of the motion principal points that were shifted from 20 to 200 pixels following a Gaussian law. Each point on the figure represents 100 trials. Figure 8.9 represents the effects of the error on the scale factor, which has been similarly set off from 2.5% to 25%. Among the numerous conclusions that can be drawn from the graphs, we would like to emphasize the following facts:

- The effects of the imprecision of intrinsic parameters are significant; however, until relatively large errors are reached (10% on the scale factors, several tens of pixels for the principal point), these effects are less significant than those due to noise (for example, if the image noise increases from 0.6 to 1.0 pixels).
- The sensitivity to errors on the principal point is less than the sensitivity to errors on the scale factor: in terms of relative errors, a 120 pixel shift of the principal point is 50% and has the same effects as a 25% error on the scale factors.
- The iterative criterion **MIN-DIST** is more sensitive to the imprecision of intrinsic parameters than the solution **FACTOR**. This can be ex-

plained by the fact that the fundamental matrix, which is directly used by **FACTOR**, partially retains the information on the exact intrinsic parameters, whereas the iterative method compensates entirely the error on the intrinsic parameters by an error on the computed motion.

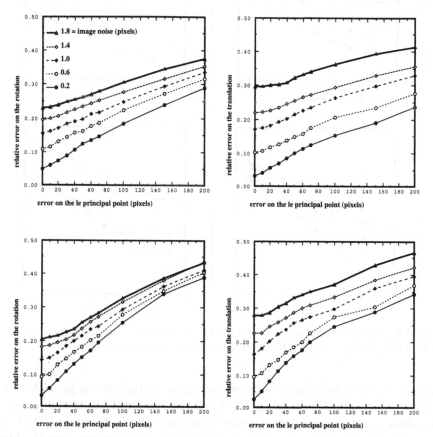

Fig. 8.8. Sensitivity of motion computation to errors on the principal point. *Top*: **FACTOR**, *Bottom*: **MIN-DIST**, *Left*: rotation, *Right*: translation.

8.5   Computing the Extrinsic Parameters of a Stereo Rig

In the usual calibration method, we work in the world coordinate system, using a 3D model of an object present in the environment. It is assumed that we know the 3D coordinates of some of its reference points, in a coordinate system attached to the object. The extrinsic parameters then consist in the displacement from the object's coordinate system (taken to be identical to the

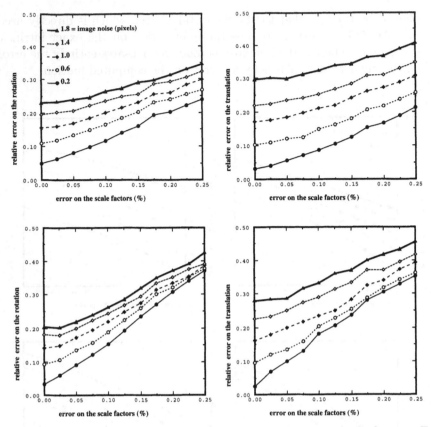

Fig. 8.9. Sensitivity of motion computation to errors on the scale factors. *Top*: **FACTOR**, *Bottom*: **MIN-DIST**, *Left*: rotation, *Right*: translation.

world reference frame) to the camera coordinate system. In the present work, we do not use any 3D model, so we do all the computations in the cameras coordinate system, and use as a reference frame the first camera. Thus, in our case, the extrinsic parameters consist in the displacement from the first to the second camera, computed in the first camera coordinate system. Since we do not have any metric information, we can compute this displacement only up to a scale factor.

Two different approaches are presented. The first one is straightforward in the case of a binocular stereo rig and more subtle in the case of a trinocular stereo rig, but it needs inter-camera point matching. The second enables us to obtain the inter-camera relative displacements using only monocular point matches. Two displacements of the stereo rig are, in general, sufficient to obtain a unique solution.

8.5.1   A Direct Approach: Binocular and Trinocular Stereo Rig

The most straightforward approach is to apply the techniques previously presented using point correspondences established between the different cameras of the stereo rig. The advantage of this method is that, since the relative displacement between the cameras is supposed to be fixed, it is possible to accumulate point matches between pairs of images taken at different times. Using multiple displacements, it is possible to obtain a number of point matches far larger than the one that could be obtained from a single pair of images, which allows one to obtain very precise results. Now let us explain how the perspective projection matrices (see Sect. 8.1.1) are obtained.

The Binocular Case. This is a very simple case. One simply uses the perspective projection matrices:

$$P_1 = [A_1, \ 0] \quad P_2 = [A_2, \ 0]D^{-1} \qquad (8.20)$$

As we know only the direction of the translation, we usually represent t as a unit vector, which allows us to perform the 3D reconstruction up to a scale factor. One piece of metric information concerning the motion, or a length measured in an image, is sufficient to obtain the scale factor.

The Trinocular Case. If we want to perform the reconstruction using three views[3] (designated by 1, 2 and 3), as the displacements D_{12} and D_{23} are known only up to a scale factor, using the formula (8.20) to obtain P_2 and P_3 yields an incorrect result, in which the epipolar constraint between the images 1 and 3 is not satisfied. The reason is that the ratio $\|t_{12}\|/\|t_{23}\|$ must be preserved, as well as the relative signs. The difficulty comes from the fact that if we know two displacements only up to a scale factor, it is only possible to determine the rotation:

$$R_{13} = R_{23}R_{12} \qquad (8.21)$$

but not the direction of the translation of $D_{12}D_{23}$, the only constraint being that it belongs to the plane $\langle t_{23}, R_{23}t_{12}\rangle$:

$$t_{13} \cdot (t_{23} \wedge R_{23}t_{12}) = 0 . \qquad (8.22)$$

In order to determine this direction, we have to know the displacement D_{13}, that is to determine it from images, which is likely to be possible, as if we want to reconstruct from the images 1, 2, and 3, there must be a portion of the scene visible in the image 1 and the image 3. Let λ, be the ratio of the norms of t_{12} and t_{23}. By expressing the proportionality constraint:

$$t_{13} \wedge (R_{23}u_1 + \lambda u_2) = 0 \qquad (8.23)$$

---

[3] It is well known that trinocular stereo algorithms are more efficient and yield more precise 3D reconstructions.

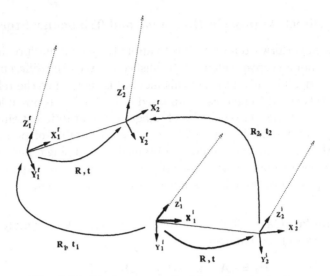

Fig. 8.10. Displacement of a binocular stereo rig.

where $u_1 = t_{12}/\|t_{12}\|$ and $u_2 = t_{23}/\|t_{23}\|$ we obtain:

$$\lambda = -\frac{(t_{23} \wedge R_{23}u_1)_1}{(t_{23} \wedge u_2)_1} = -\frac{(t_{23} \wedge R_{23}u_1)_2}{(t_{23} \wedge u_2)_2} = -\frac{(t_{23} \wedge R_{23}u_1)_3}{(t_{23} \wedge u_2)_3} . \quad (8.24)$$

Taking $t_{12} = u_1$, $t_{23} = \lambda u_2$, we then obtain, using (8.20), three perspective projection matrices that are all mutually coherent.

8.5.2   An Indirect, Monocular Approach

The Principle. In this approach, we need only to compute the displacement D_1 of the first camera and the displacement D_2 of the second camera, in the coordinate system of the first and of the second cameras, respectively. The difficulty arises from the fact that D_1 and D_2 are then known in different coordinates systems, as shown in Fig. 8.10, in which the superscripts i and f refer to initial and final positions. To cope with this problem, the idea of our method is to use the commutativity of the following diagram:

$$\begin{array}{ccc} C_1^f & \xrightarrow{\ D\ } & C_2^f \\ {\scriptstyle D_1}\uparrow & & \uparrow{\scriptstyle D_2} \\ C_1^i & \xrightarrow{\ D\ } & C_2^i \end{array}$$

where the relative displacement from the first to the second does not change, from the initial to the final position of the stereo rig, since they are rigidly attached to each other, to write the matrix equation:

$$DD_1 = D_2D \quad (8.25)$$

where D is the 4×4 unknown matrix of the displacement from the first camera to the second camera, and D_1, D_2 are the 4 × 4 displacement matrices of the first and of the second cameras, in their initial respective coordinate systems. Equation (8.25) can be decomposed into the following two matrix equations:

$$RR_1 = R_2R \qquad (8.26)$$

$$(I - R_2)t = \mu_2 t_2 - \mu_1 Rt_1 \qquad (8.27)$$

where μ_1 and μ_2 are unknown scale factors associated to D_1 and D_2, respectively. The first equation has been much studied in the framework of hand-eye calibration [38], [39], [40], [41]. Thus the reader is referred to those references for a more detailed analysis of uniqueness and sensitivity. We just show below that if we do two displacements of the stereo rig, we can solve the two resulting matrix (8.26) to compute R. The solution of the two resulting vector (8.27) to compute t up to a scale factor is less classical, since it involves working only up to a scale factor, as in the previous case of the trinocular stereo rig which it generalizes.

There is an important advantage of this method over the one which consists in computing directly the displacement from matches between the first and the second camera: since we work in each camera independently, we need only monocular matches which are more easy to obtain, as an arbitrary number of intermediate movements can be done, and a token tracking procedure used. On the opposite, finding directly stereo matches can be difficult if the baseline of the stereo rig is large, since at this stage the stereo rig is not yet calibrated.

Recovering the Rotation. To solve (8.26), we use a quaternion representation of the rotations [42]

$$q_R = (s, v)$$
$$q_{R_1} = (s_1, v_1)$$
$$q_{R_2} = (s_2, v_2) \,.$$

Writing (8.26) with this representation yields

$$q_R \times q_{R_1} - q_{R_2} \times q_R = 0 \qquad (8.28)$$

which gives the two equations:

$$s(s_1 - s_2) = v.(v_1 - v_2) \qquad (8.29)$$
$$s(v_1 - v_2) + (s_1 - s_2)v + v \times (v_1 + v_2) = 0 \,.$$

Let us write $v = \alpha v_1 + \beta v_2 + \gamma(v_1 \times v_2)$. After some algebra using the properties of the quaternions, we obtain

$$\alpha = \beta \,, \quad s_1 = s_2 \qquad (8.30)$$

and then

$$s = \gamma(v_1^2 + v_1.v_2) \tag{8.31}$$

$$\alpha^2\|v_1 + v_2\|^2 + \gamma^2\left(\|v_1 \times v_2\|^2 + \frac{\|v_1 + v_2\|^4}{4}\right) = 1. \tag{8.32}$$

Using the new coordinate system:

$$u_1 = \frac{v_1 + v_2}{\|v_1 + v_2\|}$$

$$u_2 = \frac{v_1 \times v_2}{\|v_1 \times v_2\|}$$

$$u_3 = u_1 \times u_2$$

(8.32) can be written more simply:

$$x^2 + (1 + k^2)y^2 = 0 \tag{8.33}$$

where $x = \alpha\|v_1 + v_2\|$, $y = \gamma\|v_1 \times v_2\|$, and $k = \|v_1 + v_2\|^2/2\|v_1 \times v_2\|$. This last equation determines a one-parameter family of rotations, which are parameterized by an ellipse lying in the plane (u_1, u_2).

Using two movements yields a second ellipse lying in a plane (u_1', u_2'). Some calculations show that in general the intersection of the two ellipses (8.33) is nonempty if the following condition holds:

$$k'^2(v_3.v_1')^2 = k^2(v_3'.v_1)^2. \tag{8.34}$$

This intersection yields an unique solution if the axes v_1 and v_1' (resp. v_2 and v_2') are different. If this is the case, a closed-form solution can be easily computed for the intersection of ellipses of the form (8.33), thus we can solve for α and γ, and obtain v and s using the relations (8.31) and (8.32).

We now explain why during the displacement of a stereo rig the condition (8.34) is always satisfied. Let us suppose that we do two displacements so that we have:

$$RR_1 = R_2R$$
$$RR_1' = R_2'R.$$

From these relations, we see that we have also:

$$R(R_1R_1') = (R_2R_2')R.$$

Using the constraint (8.30) on the last equation yields $s_1s_1' - u_1.u_1' = s_2s_2' - u_2.u_2'$. Since $s_1 = s_2$ and $s_1' = s_2'$, we finally obtain the equation

$$u_1.u_1' = u_2.u_2'. \tag{8.35}$$

We have then shown that this equation is equivalent to the constraint (8.34).

A Linear Method to Take into Account Multiple Motions. We can also as in [38] use the quaternion representation of rotations to obtain a linear solution. We can notice that (8.28) has the same form as (8.19). Thus there exists a 4×4 matrix G, such that:

$$q_R \times q_{R_1} - q_{R_2} \times q_R = G q_R ; \qquad (8.36)$$

G is given by a formula similar to (8.19). A closed-form solution can be obtained with two (8.36) obtained by two displacements of the stereo rig. If we use more displacements, we can improve the results by using a linear least squares procedure.

Recovering the Direction of the Translation. We suppose that we have computed R, as previously described. A geometrical analysis shows that the matrix $I - R_2$ maps all vectors in the plane perpendicular to the axis u_2 of the rotation R_2. Thus, starting from relation (8.27), we can write

$$u_2.(\mu_2 t_2 - \mu_1 R t_1) = 0 .$$

This allows us to determine the ratio $a = \mu_1/\mu_2$. It is then possible to recover the direction $t_\perp$ of the component of t orthogonal to u_2, yielding the constraint

$$t \in \langle t_\perp, u_2 \rangle . \qquad (8.37)$$

If a second movement, *for which the axis u'_2 of the rotation is different*, is used, we can compute similarly a direction $t'_\perp$. Combining the two constraints (8.37), and the same with primes, we obtain t up to a scale factor by

$$t = \lambda(t_\perp \times u_2) \times (t'_\perp \times u'_2) . \qquad (8.38)$$

Note that if we perform more than two displacements, (8.38) can be easily solved by using a linear least squares procedure. This completes the computation of the relative position of the two cameras, up to a scale factor.

8.6   Experimental Results

8.6.1   An Example of Calibration of a Binocular Stereo Rig

Self-Calibration of a Camera. We first show the results of the monocular self-calibration using three images taken by the left camera at different positions of the stereo rig. Results are quite similar for the second camera. In order to make comparisons possible with the standard calibration method, we have performed displacements in such a way that the calibration grid always remains visible in the left camera.

We use between 20 and 30 corners, which are extracted with a subpixel accuracy, semi-automatically by the program of Blaszka and Deriche [13].

Correspondence is, in this experiment, performed manually, and followed by an automatic elimination of false matches. It should be noted that the corresponding points between pairs of images are different, that is, points need not be seen in the three views. Figure 8.11 shows the points of interest matched between image 1 and image 2. The standard calibration is performed on each image, using the algorithm of Robert [20], which is a much improved version of the linear method of Faugeras and Toscani [3]. From the projection matrices obtained by this algorithm, the three fundamental matrices F_{12}, F_{23}, F_{13} are computed and used as a reference for the comparisons with our algorithm which computes the fundamental matrices from the point matches. The resulting epipoles are shown in Table 8.1. It can be seen that the estimation is quite precise. The low value of the RMS error (which represents the average distance of corresponding points to epipolar lines) confirms the validity of our linear distortion-free model, as well as the accuracy of the corner detection process. Some epipolar lines obtained with points that are seen in the three images are shown (Fig. 8.12) to illustrate the quality of the estimated epipolar geometry.

The cameras' intrinsic parameters are then computed from the fundamental matrices. We show in Table 8.2 the intrinsic parameters obtained by the standard calibration method using each of the three images, and the results of our method, with the polynomial method, and the iterative method used to compute all the parameters, or just the scale factors, starting from the previous value. It can be noted that no initial guess is required at all for the general method. The scale factors are determined with a good accuracy; however, this is not the case for the coordinates of the principal point. Thus the best is to assume that it is at the center of the image. We have then compared in Table 8.3 the camera motion obtained directly from the projection matrices given by the classic calibration procedure, and the estimation by performing the decomposition of the fundamental matrices already obtained, and using the camera parameters obtained by the self-calibration method. As the table shows the relative error on the rotation angle and the angular error on the rotation axis and translation direction, it is easy to see that the estimation is accurate.

Table 8.1. Results of the fundamental matrix estimation in the left camera.

	From the grid				Estimated				
	e_x	e_y	e'_x	e'_y	e_x	e_y	e'_x	e'_y	RMS
1–2	−222.4	181.0	466.9	167.5	−200.0	185.8	−447.5	170.1	0.36
2–3	2226.9	−1065.1	−2817.9	1646.6	2708.5	−1380.1	−2099.6	1315.5	0.31
1–3	654.4	−288.8	1114.7	−715.6	680.2	−321.7	1230.9	−842.2	0.26

Fig. 8.11. A pair of images with the detected corners superimposed.

Fig. 8.12. A triplet of images with some estimated epipolar lines superimposed.

Table 8.2. Results of the intrinsic parameters estimation in the left camera.

Method	α_u	α_v	u_0	v_0	$\theta - \frac{\pi}{2}$
grid, image 1	657.071	1003.55	244.227	256.617	$-2.05e{-}06$
grid, image 2	664.975	1015.2	232.61	257.701	$-7.47e{-}07$
grid, image 3	639.749	980.185	252.174	249.585	$-2.60e{-}06$
Kruppa polynomial	639.405	982.903	258.980	341.013	$-6.11e{-}03$
Kruppa iterative	640.12	936.08	206.17	284.95	-0.07
Kruppa iterative (center)	681.28	985.69	255	255	0

Table 8.3. Results of the camera motion estimation in the left camera (first sequence).

Movement	r_x	r_y	r_z	t_x	t_y	t_z	$\frac{\Delta\alpha}{\alpha}$	θ_r	θ_t
1–2 grid	0.01175	−0.2117	−0.01785	−0.7290	−0.06831	0.6809			
estimated	0.01843	−0.2110	−0.01961	−0.7239	−0.06102	0.6871	0.0005	1.8	0.62
2–3 grid	0.1900	0.4526	0.1211	−0.9395	0.2779	0.1999			
estimated	0.1915	0.4682	0.1279	−0.9209	0.2896	0.2608	0.032	0.61	3.7
1–3 grid	0.2007	0.2533	0.07876	0.6976	−0.5041	0.5090			
estimated	0.01306	−0.2145	−0.01405	−0.7371	−0.05872	0.6731	0.10	0.98	3.0

Extrinsic parameters computation Once the self-calibration of each camera has been achieved, we have performed two other displacements of the stereo rig. We have not used the three previous displacements because they yield computations that are less stable for the method we want to illustrate now: the computation of the relative displacement between the two cameras of the rig using only monocular matches, the problem being the difference of motion between the two cameras of the rig. We have performed small displacements which maximize this difference. The six images are shown in Fig. 8.13. Since only a small part of the calibration grid is seen, we cannot directly check the results of the determination of camera motion shown in Table 8.4. However, we verify the consistency of these results thanks to two families of constraints: one arising from the fact that the two cameras of the rig are rigidly attached, and one arising from the fact that the third displacement is a composition of the first two displacements, as only three *images* are used. The binocular constraints are that the angles of rotation of the two cameras are equal for a given displacement of the rig (8.30), which can be checked in Table 8.4 and the relation (8.35), whose residual values are here −0.00943, .01950, and −0.05394. The monocular constraints are obtained from the fact that the composition of the two first motions gives the third one. We obtain for the rotations, using (8.21):

$$r_1'' = [-0.01125, 0.2027, -0.1389]^{\mathrm{T}} \quad r_2'' = [-0.02648, 0.1936, -0.1518]^{\mathrm{T}}$$

which is close to the values actually computed and shown in Table 8.4: the relative error on the angles is 1% and 0.8%, and the angle between the axes is 8.7° and 2.9°. The mixed product (8.22) involves also the direction of translations. The value in the left camera is −0.0192, and in the second camera −0.002. Thus we have checked that all the constraints are well satisfied.

We then computed the relative displacement between the two cameras of the rig, using different methods:

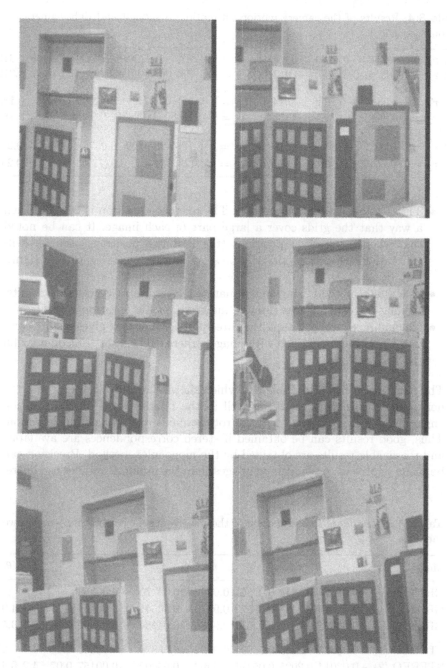

Fig. 8.13. Three pairs of stereo images (disposed for cross-viewing).

Table 8.4. Results of the camera motion estimation in the left and right camera (second sequence).

1–2	$r_1 = [-0.00012, 0.3130, 0.00773]^{\mathrm{T}}$	$t_1 = [0.1237, -0.0209, 0.9920]^{\mathrm{T}}$	$\alpha_1 = 0.3131$
	$r_2 = [0.00554, 0.31196, -0.01219]^{\mathrm{T}}$	$t_2 = [0.2953, 0.0160\ 0.9552]^{\mathrm{T}}$	$\alpha_2 = 0.3122$
2–3	$r'_1 = [-0.0334, -0.1098, -0.143]^{\mathrm{T}}$	$t'_1 = [-0.124, 0.5974, 0.7922]^{\mathrm{T}}$	$\alpha'_1 = 0.1833$
	$r'_2 = [-0.0540, -0.117, -0.133]^{\mathrm{T}}$	$t'_2 = [-0.01089, 0.02439, 0.9996]^{\mathrm{T}}$	$\alpha'_2 = 0.1853$
1–3	$r''_1 = [-0.00224, 0.2054, -0.1307]^{\mathrm{T}}$	$t''_1 = [-0.1882, 0.9809, 0.0476]^{\mathrm{T}}$	$\alpha''_1 = 0.2435$
	$r''_2 = [-0.06175, 0.198, -0.1385]^{\mathrm{T}}$	$t''_2 = [0.3423, 0.1487, 0.9277]^{\mathrm{T}}$	$\alpha''_2 = 0.2494$

- The classical calibration method. The reference position is taken in such a way that the grids cover a large part of each image. It can be noted that when using other positions (the first two positions used for self-calibration, where the grids can be seen entirely), the results vary significantly.
- The direct method using stereo matches. This yields very stable results. Adding correspondences through images improves rotation accuracy.
- The indirect method, using the three pairs of motions, gives results comparable to those obtained with images where calibration grids do not "fill the image frame".

The results are shown in Table 8.5, which shows the rotation vector and the normalized translation vector, as well as the relative error on the rotation angle and the angular error and the rotation axis and translation direction. Thus, good results can be obtained if stereo correspondences are available, and reasonable results are obtained by the monocular method. Precision can be easily improved by using more images than the minimal number used here.

Table 8.5. Results of the estimation of the relative displacement between the two cameras.

Method	r_x	r_y	r_z	t_x	t_y	t_z	$\frac{\Delta\alpha}{\alpha}$	θ_r	θ_t
GRID (ref.)	−0.04097	0.1842	0.05561	0.9992	0.03770	−0.00889			
GRID (1)	−0.04015	0.2285	0.05573	0.9970	0.04541	0.0613	0.21	3.8	4.0
GRID (2)	−0.03595	0.2042	0.05611	0.9976	0.04234	0.05303	0.09	2.7	3.5
STEREO (1)	−0.03383	0.2205	0.05298	0.9992	0.03879	−0.00402	0.16	4.7	0.28
STEREO (2)	−0.03014	0.2025	0.05411	0.9992	0.03902	−0.00157	0.07	4.2	0.42
STEREO (1 + 2)	−0.04307	0.1895	0.05411	0.9991	0.03804	−0.01655	0.025	0.9	0.44
MONO	−0.04915	0.2383	0.05322	0.9987	−0.002223	−0.04902	0.26	4.1	3.2

8.6.2 Reconstructions from a Triplet of Uncalibrated Images Taken by a Camera

We now show an example of reconstruction obtained in the more general case of structure from motion. We use only three uncalibrated views taken by the same camera, shown in Fig. 8.14. Edge detection is performed. Then the edge chains are approximated by B-splines, which are given as input to the trinocular stereovision algorithm of Robert [43,20]. The matching phase of this algorithm uses only the epipolar geometry obtained from the fundamental matrices F_{12}, F_{13}, and F_{23}, which are computed from the point correspondences. The 3D reconstruction phase requires in addition three projection matrices which relate the three image coordinate systems to a common world coordinate system. They are obtained by taking as the world coordinate system, the first camera coordinate system, and by finding the two displacements D_{12}, D_{23}, as well as the ratio of the norms of t_{12} and t_{23} (for which the computation of D_{13} is needed). Results of reconstruction are shown in Fig. 8.15

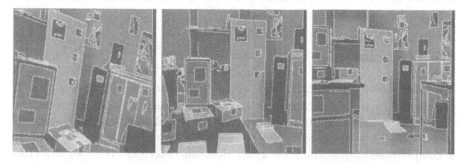

Fig. 8.14. The triplet of images, with edge chains superimposed.

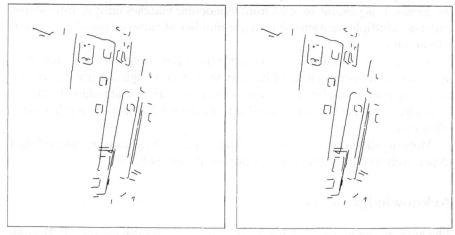

Fig. 8.15. Reconstruction from the uncalibrated triplet (stereogram for cross-viewing).

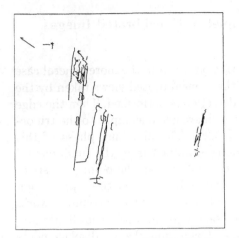

Fig. 8.16. Two reprojected views.

as a stereogram; planar structures and angles are well-preserved. It can be seen in Fig. 8.16, which represents two reprojected views, that the distances estimated are also metrically correct.

8.7   Conclusion

We have shown theoretically and with experiments on real images that it is possible to calibrate completely a stereo rig just by pointing it at the environment, selecting points of interest, and tracking them independently in each image while moving the stereo rig with an unknown motion. By a complete self-calibration, we mean that we determine the two cameras' intrinsic parameters and the relative displacement between the two cameras. All these parameters are computed without using any a priori knowledge of the scenes. It is possible to start from monocular matches only, or from stereo matches. Multiple positions or multiple number of cameras can also be taken into account.

The present limitation of the methods comes from the fact that it is necessary to measure points of interest with a very high accuracy if we are to obtain results similar to the ones obtained with the standard methods. However, we think that our method will prove useful, due to its much greater versatility.

More details on the Fundamental Matrix and Self-calibration can be found respectively in the subsequently published papers [49], [50].

Acknowledgements

The authors would like to thank R. Deriche for useful discussions, T. Blaszka for providing us with the corner detection program, L. Robert for helping us with his calibration and stereo software, and H. Mathieu for connecting the cameras.

References

1. J. L. Mundy, A. Zisserman (eds.) *Geometric invariance in computer vision.* MIT Press, 1992.
2. J.G. Semple, G.T. Kneebone. *Algebraic projective geometry.* Oxford Science Publication, 1952.
3. O.D. Faugeras, G. Toscani. The calibration problem for stereo. In *Proceedings of CVPR'86*, pp. 15–20, 1986.
4. R.Y. Tsai. An Efficient and Accurate Camera Calibration Technique for 3D Machine Vision. In *Proceedings CVPR '86, Miami Beach, Florida*, 364–374. IEEE, June 1986.
5. H. Asada, M. Brady. The curvature primal sketch. IEEE Transactions on Pattern Analysis and Machine Intelligence, **8**, 2–14, 1986.
6. G. Medioni. Y. Yasumuto. Corner detection and curve representation using cubic b-spline. In *Proc. International Conference on Robotics and Automation*, pp. 764–769, San Francisco, 1986. IEEE.
7. F. Veillon, R. Horaud, T. Skordas. Finding geometric and relational structures in an image. In O. Faugeras (ed.) *Computer Vision – ECCV'90*. (Lecture Notes in Computer Science, Vol. 427) pp. 374–384, Springer, Berlin Heidelberg, 1990.
8. M.A. Shah, R. Jain. Detecting time-varying corners. Computer Vision, Graphics, and Image Processing **28**, 345–355, 1984.
9. C. Harris, M. Stephens. A combined corner and edge detector. In *Proc. Alvey Vision Conference*, 189–192, Manchester, 1988.
10. R. Deriche, G. Giraudon. A Computational Approach for Corner and Vertex Detection. *The International Journal of Computer Vision* **10(2)**, 101–124, 1993.
11. A. Guiducci. Corner characterization by differential geometry techniques. Pattern Recognition Letters **8**, 311–318, 1988.
12. K. Rohr. Modelling and identification of characteristic intensity variations. Image and Vision Computing **10(2)**, 66–76, 1992.
13. R. Deriche and T. Blaszka. Recovering and characterizing image features using an efficient model based approach. In *Proc. International Conference on Computer Vision and Pattern Recognition*, pp. 530–535, New York, 1993. IEEE.
14. R.E. Kelly, P.R.H. McConnell, S.J. Mildenberger. The gestalt photomapper. Photogrammetric Engineering and Remote Sensing **43**, 1407–1417, 1977.
15. W. Forstner, A. Pertl. Photogrammetric standard methods and digital image matching techniques for high precision surface measurements. In Gelsema, E.S., Kanal, L.N., (eds.), *Pattern Recognition in Practice II*, pp. 57–72. Elsevier Science Publishers, 1986.
16. D.B. Gennery. *Modelling the Environment of an Exploring Vehicle by means of Stereo Vision.* PhD thesis, Stanford University, June 1980.
17. R. Deriche, O.D. Faugeras. Tracking line segments. Image and Vision Computing, **8(4)**, 261–270, 1990. A shorter version appeared in the Proceedings of the 1st ECCV.

18. O.D. Faugeras. What can be seen in three dimensions with an uncalibrated stereo rig. In G. Sandini (ed.) *Computer Vision - ECCV'92*. (Lecture Notes in Computer Science, Vol. 588) pp. 563–578, Springer, Berlin Heidelberg, 1992.

19. R. Hartley, R. Gupta, T. Chang. Stereo from uncalibrated cameras. In *Proc. of the Conference on Computer Vision and Pattern Recognition*, pp. 761–764, Urbana, 1992.

20. L. Robert. *Reconstruction de courbes et de surfaces par vision stéréoscopique. Applications a la robotique mobile*. PhD thesis, Ecole Polytechnique, 1993.

21. A. Shashua. Projective structure from two uncalibrated images: structure from motion and recognition. Technical Report A.I. Memo No. 1363, MIT, Sept 1992.

22. H.C. Longuet-Higgins. A Computer Algorithm for Reconstructing a Scene from Two Projections. Nature **293**, 133–135, 1981.

23. Q.-T. Luong, R. Deriche, O.D. Faugeras, T. Papadopoulo. On determining the Fundamental matrix: analysis of different methods and experimental results. Technical Report 1894, INRIA, 1993.

24. E. Kruppa. Zur Ermittlung eines Objektes aus zwei Perspektiven mit innerer Orientierung. Sitz.-Ber. Akad. Wiss., Wien, math. naturw. Kl., Abt. IIa. **122**, 1939–1948, 1913.

25. O.D. Faugeras, S.J. Maybank. Motion from point matches: multiplicity of solutions. The International Journal of Computer Vision, **4(3)**, 225–246, 1990. also INRIA Tech. Report 1157.

26. S.J. Maybank, O.D. Faugeras. A Theory of Self-Calibration of a Moving Camera. The International Journal of Computer Vision, **8(2)**, 123–151, 1992.

27. O.D. Faugeras, Q.-T. Luong, and S.J. Maybank. Camera self-calibration: theory and experiments. In G. Sandini (ed.) *Computer Vision - ECCV'92*. (Lecture Notes in Computer Science, Vol. 588) pp. 563–578, Springer, Berlin Heidelberg, 1992.

28. A. Morgan. *Solving polynomial systems using continuation for engineering and science problems*. Prentice-Hall, 1987.

29. Q.-T. Luong, O.D. Faugeras. Self-calibration of a camera using multiples images. In *Proc. International Conference on Pattern Recognition*, pp. 9–12, Den Haag, 1992. IEEE.

30. Q.-T. Luong. *Matrice fondamentale et calibration visuelle sur l'environnement: vers une plus grande autonomie des systèmes robotiques*. PhD thesis, Université de Paris-Sud., Dec. 1992.

31. O.D. Faugeras, F. Lustman, G. Toscani. Motion and Structure from point and line matches. In *Proc. International Conference on Computer Vision*, pp. 25–34, London, June 1987. IEEE

32. M.E. Spetsakis, J. Aloimonos. Optimal computing of structure from motion using point correspondences in two frames. In *Proc. International Conference on Computer Vision*, pp. 449–453, Tarpen Springs, FL, 1988. IEEE.

33. J. Weng, N. Ahuja, T.S. Huang. Optimal motion and structure estimation. In *Proc. International Conference on Computer Vision and Pattern Recognition*, pp. 144–152, San Diego, 1989. IEEE.

34. B.K.P. Horn. Relative orientation. The International Journal of Computer Vision, **4(1)**, 59–78, 1990.

35. R.Y. Tsai, T.S. Huang. Uniqueness and estimation of three-dimensional motion parameters of rigid objects wirth curved surfaces. IEEE Transactions on Pattern Analysis and Machine Intelligence, **6**, 13–27, 1984.

36. R.I. Hartley. Estimation of relative camera positions for uncalibrated cameras. In G. Sandini (ed.) *Computer Vision – ECCV'92*. (Lecture Notes in Computer Science, Vol. 588) pp. 563–578, Springer, Berlin Heidelberg, 1992.

37. R. Kumar and A. Hanson. Sensibility of the pose refinement problem to accurate estimation of camera parameters. In *Proceedings of the International Conference on Computer Vision*, pp. 365–369, Osaka, Japan, 1990.

38. J.C.K. Chou and M. Kamel. Quaternions approach to solve the kinematic equation of rotation, $A_a A_x = A_x A_a$, of a sensor-mounted robotic manipulator. In *Proc. International Conference on Robotics and Automation*, pp. 656–662, Philadelphia, 1988. IEEE.

39. Y.S. Shiu, S. Ahmad. Calibration of wrist-mounted robotic sensors by solving homogeneous transform equations of the form $AX = XB$. IEEE Transactions on robotics and automation, **5(1)**, 16–29, 1989.

40. R.Y. Tsai, R.K. Lenz. Real time versatile robotics hamd/eye calibration using 3D machine vision. In *Proc. International Conference on Robotics and Automation*, pp. 554–561, Philadelphia, 1988. IEEE.

41. H.H. Chen. A screw motion approach to uniqueness analysis of head-eye geometry. In *Proc. of the Conference on Computer Vision and Pattern Recognition*, pp. 145–151, Maui, Hawaii, June 1991. IEEE.

42. E. Pervin, J.A. Webb. Quaternions in computer vision and robotics. In *Proc. International Conference on Computer Vision and Pattern Recognition*, pp. 382–383, Arlington, VA, 1983. IEEE.

43. L. Robert, O.D. Faugeras. Curve-Based Stereo: Figural Continuity And Curvature. In *IEEE Proc. International Conference on Computer Vision and Pattern Recognition*, pp. 57–62, Maui, Hawaii, June 1991. IEEE.

44. P.R. Beaudet. Rotational invariant image operators. In *Proc. International Conference on Pattern Recognition*, pp. 579–583, Kyoto, 1978. IEEE.

45. L. Dreschler, H.H. Nagel. On the selection of critical points and local curvature extrema of region boundaries for interframe matching. In *Proc. International Conference on Pattern Recognition*, pp. 542–544, Munich, 1982. IEEE.

46. L. Kitchen, A. Rosenfeld. Gray-level corner detection. Pattern Recognition Letters 95–102, 1982.

47. H.H. Nagel. Constraints for the estimation of displacement vector fields from image sequences. In *Proc. International Joint Conference on Artificial Intelligence*, pp. 156–160, Karlsruhe, 1983. Morgan Kaufman.

48. O.A. Zuniga, R.M. Haralick. Corner detection using the facet model. In *Proc. International Conference on Computer Vision and Pattern Recognition*, pp. 30–37, Arlington, VA, 1983. IEEE.

49. Q.-T. Luong, O.D. Faugeras. The fundamental matrix: theory, algorithms, and stability analysis. *The International Journal of Computer Vision*, **17(1)**, 43–76, 1996.

50. Q.-T. Luong, O.D. Faugeras. Self-calibration of a moving camera from point correspondences and fundamental matrices. *The International Journal of Computer Vision*, **22(3)**, 261–289, 1997.

Index

Springer Series in Information Sciences

Editors: Thomas S. Huang Teuvo Kohonen Manfred R. Schroeder